FOREWORD

Major changes are being proposed for the loading and design concepts of electrical transmission structures. This seminar, Innovations in the Design of Electrical Transmission Structures, was developed to review the proposed changes, present new concepts, and provide an opportunity for discussion of specific ideas as they apply to present procedures. Six formal sessions were presented covering weather-element loading conditions, structural options, design recommendations, foundation alternatives, research efforts, and code requirements.

The speakers then participated in informal question and answer sessions to discuss specific questions from the participants. The pertinent points covered in these question and answer sessions are also included in this document.

Each paper was reviewed by the coordinator for the session as well as the Proceedings Editor. All papers are eligible for discussion in the Journal of Structural Engineering and are eligible for ASCE awards.

Gene M. Wilhoite, Editor

ACKNOWLEDGEMENTS

This seminar was made possible by the joint efforts of the University of Missouri-Kansas City, the Electric Power Research Institute, the Institute of Electrical and Electronics Engineers, and the American Society of Civil Engineers. The following control people developed the format and coordinated the presentations: Dan E. Jackman, John W. Harrison, Richard E. Kennon, A.M. DiGioia, Frank A. Denbrock, and Gene M. Wilhoite. Steve Harrison, Steve Powell, and Mike Collins served as the recorders for the Discussion Sessions.

The assistance and guidance of Elizabeth Yee, Terry Williams, and Shiela Menaker of the ASCE Staff are deeply appreciated. The Executive Committee of the Structural Division of ASCE authorized this seminar and provided support for the sessions.

CONTENTS

Session I – Determination of Climatic Design Loads

LRFD Format for Transmission Structures
    Alain H. Peyrot 1

Wind Loads on Electrical Transmission Structures
    Kishor C. Mehta 11

A Field Study of Wind-Induced Conductor Loads
    Christopher Y. Tuan, Michael T. Potter, and
    Dan E. Jackman 21

Session II – Design Philosophy for Structures

Steel Poles – Past, Present, and Future
    Ronald E. Randle 30

Lattice Tower Design Using Cold-Formed Shapes
    Paolo Faggiano 37

Innovations in Prestressed Concrete Structures
    William M. Howard 47

Session III – Procedures for Member and Connection Design

Cold-Formed Angles for Transmission Towers
    Edwin H. Gaylord 57

Design and Test of Cold-Formed Members
    Adolfo Zavelani 69

Bolted Connections for Steel Towers
    Gene M. Wilhoite 79

Session IV – Foundation Design Techniques

Design of Drilled Piers Subjected to High Overturning Moments
    Anthony M. DiGioia, Jr. 86

IEEE/ASCE Transmission Structure Foundation Design Guide
    Paul A. Tedesco 96

Session V – Current Research Efforts

TLMRF Research Initiatives
    Robert A. LeMaster 106

Strain Gaging and Data Acquisition at the TLMRF
Fred Arnold 116

Strength Data Base for LRFD of Transmission Lines
Alain H. Peyrot 126

Session VI – National Electrical Safety Code Requirements

NESC – A Flexible Document
Richard A. Kravitz 135

NESC Loading Requirements
Jerome G. Hanson................................ 138

Discussion on Session I – Determination of Climatic Design
Loads 141

Discussion on Session II – Design Philosophy for Structures 144

Discussion on Session III – Procedures for Member and
Connection Design 150

Discussion on Session IV – Foundation Design Techniques....... 153

Discussion on Session V – Current Research Efforts 155

Discussion on Session VI – National Electrical Safety Code
Requirements.. 158

Subject Index 161

Author Index... 163

LRFD Format for Transmission Structures

Alain H. Peyrot[1], M. ASCE

ABSTRACT

The structural design of high voltage electric transmission lines in the US is usually governed by the National Electrical Safety Code (NESC). A simple reliability-based Load and Resistance Factor Design (LRFD) format has been proposed as an alternative or a supplement to the NESC. With the proposed LRFD method, the designer is able to assign relative levels of reliability to different lines, to different structures within one line and to different components within one structure. This paper demonstrates how the LRFD method of design is to be used and it discusses its advantages over more traditional methods.

I INTRODUCTION

Prevailing U.S. practice requires that transmission lines be designed, as a minimum, to meet the requirements of recent editions of the National Electrical Safety Code (9). The NESC rules, usually complemented by individual utility agenda (3), are largely based on successful field experience. Unfortunately, when using the NESC or other experience-based criteria, there is no yardstick to determine whether these criteria produce safe and economically successful designs.

An overdesigned line (reliability too high) has very few structural failures during its lifetime, but the unnecessary capital invested can be viewed as economic failure. On the other hand, an underdesigned line (reliability too low) experiences too many structural failures, the cost of which is unacceptable. The optimum balance between costs and risks cannot be achieved without consideration of the reliability of specific components in the line. A practical design procedure that allows the designer to select the reliability of any component in a line has been developed (Refs. 7 and 10). The procedure uses the Load and Resistance Factor Design format (abbreviated herein as LRFD). That format is fast becoming the standard structural design procedure throughout the world for all types of structures and materials (bridges, buildings, dams, etc.). The LRFD method discussed herein is reliability-based in that the designer can select the load and strength factors to force any reliability on any line component. It is one of several recently proposed methods (13).

----

1. Professor, Department of Civil and Environmental Engineering, University of Wisconsin, Madison, 53706

This paper reviews the LRFD procedure and emphasizes its practical use. While it may appear that the data that are needed to implement LRFD are not always available, it is emphasized that the use of LRFD with limited data augmented by professional judgement is an improvement over currently used methods.

II SYSTEM, SUBSYSTEMS AND COMPONENTS

The LRFD allows the designer to view an entire transmission line as a <u>system</u> consisting of several <u>subsystems</u>, each subsystems consisting of <u>components</u>. The distinction between line (the system), subsystems and components is essential if the full benefits of LRFD are to be realized. The conductors, the ground wires, the insulators, a tangent structure, an angle structure, a dead-end structure, etc., can be considered each a subsystem. A subsystem in turn is made up of components. For example, a steel tower is made up of many components: steel angle members, gusset plates, bolts and foundations. A metal H-frame is also made up of many components: cross arm, cross- and knee-braces, segments of poles (considering each 5 ft segment along each pole as a component), etc.

There are many theories that could be used to relate the reliabilities of components in a subsystem to that of the subsystem, and relate the reliabilities of the subsystems to that of the entire system. However, because of: 1) complexities of the theories, 2) occurences of loads redistribution following an initial failure and 3) lack of cross correlated strength data among components of a single subsystem, it is unrealistic to expect that accurate relationships between component reliabilities and subsystems reliabilities will ever be developed. Fortunately, it is quite reasonable to assume that the reliability of a subsystem is the same as that of its components with the smallest reliability. This is equivalent to saying that a subsystem is of the "weakest link or series" type, i.e. that it fails if any of its components fails. For example, a lattice tower can be said to have failed if any of its member has failed. It is not totally true in the case of an arm member failure (that failure would not cause the failure of the entire tower), but it is true in the case of the failure of a leg, a critical bracing member or a foundation.

Therefore, because of overwhelming practical considerations (and also because reasonable accuracy is still achieved), the <u>reliability of a subsystem</u> should be considered synonymous to the <u>reliability of its weakest component or components</u>. Since LRFD can be used to control the reliabilities of components, it can be used as well to control the reliabilities of subsystems.

In order to achieve economy in a line, the designer may assign different reliabilities to each subsystem, i.e. he/she may decide that the conductor subsystem should be 10 times more reliable than a dead-end structure which in turn should be 5 times more reliable than a tangent structure. By assigning different reliabilities to different subsystems, a preferred sequence of failure among the subsystems can be built-in.

TRANSMISSION STRUCTURES LRFD FORMAT   3

III RELIABILITY, SECURITY AND SAFETY

The nature of the events which produce loads in a transmission line may be used to classify these loads into three broad categories. Category I includes loads produced by climatic phenomena which can be described by probability laws. Loads produced by extreme winds and, where data exist, loads produced by severe combinations of ice-and-wind fall into Category I. Category II includes loads produced by natural or accidental events which, because of lack of data, cannot be described statistically. Examples of loads in Category II are loads produced by unbalanced ice or loads produced by failures of structures or components from defects, wear, fatigue, landslides, earthquakes, sabotage, etc... Finally, Category III includes loads produced by construction or maintenance operations. Like the loads in Category II, the loads in Category III cannot be described by a probability law.

Therefore, out of the many load cases normally considered in design, only those in Category I (i.e. loads produced by extreme wind and loads produced by the combination of ice-and-wind) can be considered in a formal RELIABILITY-based design procedure. The ability of a line to withstand the loads in Category II or to limit damage from these loads to adjacent structures is a measure of the line SECURITY. Designing security in a line involves selecting appropriate design loads and structure configurations to preclude the possibility of cascading. The ability of a line to safely support the loads in Category III is a measure of the line SAFETY.

IV LOADS AND STRENGTHS VARIABILITY

The actual extreme wind or ice events which will affect a transmission line during any one year period (or during its planned lifetime) are not known. They can only be described by their Probability Density Functions (PDF's) as shown in Fig. 1. It is commonly assumed that these PDF's are of the Extreme I (Gumbel) type. Wind and ice data can be collected to determine the two parameters (say mean, m, and coefficient of variation, V) of these distributions. Twenty to thirty years of record are considered sufficient for good accuracy, but shorter records can be used (5). Once a PDF is established, values of the variable with given return periods, RP (in year), can be determined. A value with a RP-year return period has probability of 1/RP of being exceeded in one year. The wind velocity that has a 50-year return period, W_{50}, is the basic design wind velocity value used by the NESC as well as the LRFD. It is the same design variable used in the most recent US and foreign building codes. While the NESC specifies arbitrary combinations of ice and wind, the LFRD ice-and-wind case is consistent with the extreme wind case in that it uses a basic design ice thickness, I_{50}, with a 50-year return period. Contour maps giving approximate values of W_{50} and I_{50} for wide areas of the US are included in Ref. 7.

The actual strength, R, of any transmission structure component is also a random variable that can be characterized by a PDF as shown in Fig.2. It is commonly assumed that the PDF is Normal (Gaussian). The PDF is entirely known if two of its parameters are known, say its mean, m_R, and its coefficient of variation, V_R. The nominal strength R_n used in

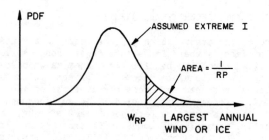

Figure 1. Probability Density Function for climatic events

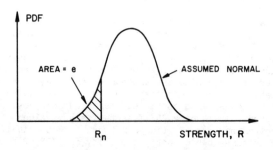

Figure 2. Probability Density Function for component strength

connection with the NESC or the LRFD is a predictor of the actual strength of the component. Several design guides have become de facto standards for calculating R_n for typical components (1,2,4,6,12). Unfortunately, none of the guides indicate how R_n is statistically related to R. Recent research efforts have attempted to estimate m_R and V_R. The Transmission Line Mechanical Research Data Base (TLMRDB) developed by the Electric Power Research Institute (EPRI) will be an invaluable source of data over the next few years (11). From m_R and V_R the probability that R is below R_n can be calculated. If that probability is "e" percent, R_n is said to be the "e" percent lower exclusion limit of strength. The notation R_e is used in this paper to indicate a strength at the "e" percent exclusion limit.

The exclusion limits of the strengths R_n currently used (1,2,4,6,12) vary widely. The exclusion limit for the strength of a steel lattice tower member based on the strength formulae in Ref. 6 and on real test data from many references is in the range of 5 to 15 percent. While the variety of exclusion limits inherent in current strength calculation procedures is not an obstacle to the production of reliability-consistent designs with the proposed LRFD method, it is recommended by the author that <u>all future editions of component strength guides use a uniform 5 percent exclusion limit</u>. The 5 percent exclusion limit is low enough to be chosen as the value at which the component can be proof loaded.

Typical transmission structure components have coefficients of variation in the range of 5 to 20 percent. Wood components, especially components which have been in service for long periods of times, may have coefficients of variation in excess of 20 percent.

V TARGET RELIABILITY

At the planning stage of a line the engineer should decide on target reliability levels (or target probabilities of failures) for each subsystem and each component within a subsystem. Since reliability has to be defined for a fixed time period, all reliability values described herein are based on a one year line exposure to climatic events.

It is convenient to express the target annual probability of failure of a component P_f by the equation

$$P_f = P_0 / (G C) \qquad (1)$$

where P_0 is the base **annual** probability of failure defined in details in Ref. 10, where the global importance factor G allows adjustement of the reliability of all the components of the line or a particular subsystem, and where the component factor C allows further adjustment of the reliability of a particular component within a subsystem.

The designer should not be concerned by the exact numerical value of P_0 (about .2 percent). P_0 is a standard value that will apply to all transmission lines. Instead, he/she should decide on what additional (or reduced) reliability is needed over the standard design, i.e. what factors G and C should be used. With the proposed scheme, the product GxC represents, on a relative scale, the reliability of any component in the line.

The ability to assign relative reliability values to various lines, to various structures within one line, and to various components within one structure is an extremely useful concept. For example, a utility may require a five-fold increase of reliability (G = 5) for an essential line and another five-fold increase (G = 25) for the dead ends in that line. The utility may also want to design a preferred sequence of failures in components of a given structure if loads in excess of the design loads occur. This can be done by assigning lower reliabilities to the components which should fail first. The utility may also compare the cost of an item to the consequences of the failure of that item. If the cost is low and the consequences very costly, a higher reliability (say C = 10) may be required for that component. The reliability of a component for which C = 10 and G = 25 is therefore 250 times larger than that of a component for which both C and G are equal to one.

Use of the global importance factor is equivalent to assigning a class to a line. The IEC (8) has recommended that 3 security classes of lines, labeled 1, 2 and 3, be considered and that for each class the design loads be based on a 50-, 100-, and 500-year return period, respectively. From results in Ref. 10, it can be shown that the 3 IEC classes are equivalent to assigning global importance factors G =1 , 2 and 10. Class 3 is therefore 10 times more reliable than class 1.

The power of the LRFD method is that the target reliability of a component (represented by both factors G and C) can be achieved by simply selecting a load factor corresponding to G and a strength factor corresponding to C as described next.

VI LRFD FORMAT

The following are the LRFD design equations:

$$\phi R_n > \text{Effect of } [\, \gamma_D\, DL + \gamma_W\, WL_{50}\,] \qquad (2)$$

$$\phi R_n > \text{Effect of } [\, \gamma_D\, DL + \gamma_I\, IL_{50} + \gamma_W\, WIL_r\,] \qquad (3)$$

$$\phi R_n > \text{Effect of } [\, \gamma_D\, DL + (\, AL \text{ or } CL\,)\,] \qquad (4)$$

where:

ϕ = a strength (or resistance, or capacity) reduction factor which is used to adjust the reliability of a component within a subsystem. ϕ can be read off directly from Table 1 as a function of the component factor C. ϕ takes into account variabilities in material, dimensions, and workmanship, and the uncertainty inherent in the equation used to calculate R_n as a predictor of the true strength of the component

Table 1. Strength factor ϕ to adjust component reliability by factor C e = exclusion limit of R_n

C =	50	10	2	1	1/3
e = 1 pct	.65	.89	1.04	1.16	1.39
5	.56	.77	.90	1.00	1.20
10	.52	.72	.84	.93	1.12
15	.50	.69	.80	.89	1.07

Table 2. Load factor γ_W to adjust component reliability by factor G Load factor γ_I = 1.2 times the values in this table

G =	1	2	3	4	6	9	13	19	28	40	56	80	110
	1.0	1.1	1.2	1.3	1.4	1.5	1.6	1.7	1.8	1.9	2.0	2.1	2.2

R_n = the nominal strength of the component. The proposed LRFD procedure requires that the exclusion limit of R_n be known or estimated

γ_D = the load factor applied to the dead loads. A value of 1.2 was recommended in Refs. 7 and 10. A value of 1.4 is better for cases where the dead load in a component exceeds one half of its total load.

DL = the dead loads

γ_W = the load factor applied to the wind loads WL_{50} or to the wind loads WIL_n. The load factors γ_W and γ_I which account for the importance of the line or the importance of a particular subsystem in the line are obtained from Table 2

WL_{50} = loads produced by the fastest mile wind velocity W_{50} which has a 50-year return period. W_{50} may be determined from local meteorological data or it may be obtained from a regional map (7). The wind direction is taken as that which produces the largest load effect. The loads WL_{50} are loads on the wires or loads acting directly on the supporting structures. The loads WL_{50} together with temperature affect wire tensions. The temperature at which tensions should be calculated is that most likely to occur at the time of the extreme wind event. The tensions, in turn, affect the transverse forces on an angle structure and the transverse and longitudinal forces on a dead end structure

γ_I = the load factor applied to the ice loads IL_{50} (see Table 2)

IL_{50} = vertical loads produced by the ice thickness I_{50} which has a 50-year return period. The ice should be assumed uniformly deposited around all the wires. The ice loads produced by ice accumulation on the supporting structure itself should also be considered if they are significant

WIL_r = loads produced by the reduced wind velocity W_r blowing on wires and components coated with the ice thickness I_{50}. These loads may be called wind-on-ice loads. W_r is the largest wind velocity which is expected to occur over the duration of the icing event. W_r should be applied in the direction which results in the largest load effect and the tensions produced by the ice-and-wind event should be determined at the temperature most likely to occur during that event (see Ref. 7 for suggested numerical values)

AL = loads produced by assumed accidental events
(loads in Category II)

CL = loads produced by construction/maintenance operations
(loads in Category III)

The load factors γ_W and γ_I in Eqs. 2 and 3 are applied to the forces created by the wind or the ice-and-wind on the wires (see Fig.3) and to the forces produced by the same wind or the ice-and-wind directly on the supporting structure. The procedure is quite different from that required by the NESC, whereby the load factors are applied to the loads transferred from the wires to the structures as shown in Fig. 4 for the ice-and-wind case. Therefore, unlike the NESC, the proposed LRFD method does not discriminate between the directions (vertical, transverse and longitudinal) of the resultant loads; rather it considers the load source. The LRFD procedure guaranties that nonlinear effects do not affect the line calculated reliability.

While it is possible to vary the reliability of a component by varying only the strength factor or varying only the load factors, there are theoretical and practical advantages to using both strength and load factors. The strength factor should be used to fine tune the relative reliabilities of components within one subsystem (represented by the factor C) while the load factors are used to control the overall reliability of a line or the relative reliabilities of different subsystems within one line (represented by the factor G). The practical reason for the last recommendation is that only one set of design loads needs to be developed for a specific subsystem, even if different components within that subsystem are assigned different reliability levels. Having only one set of loads and consequently one set of load effects to consider is very important because the LRFD method applies the load factors at the wire level. Additional sets of loads would require additional sets of calculations for wire tensions and additional structural analyses.

TRANSMISSION STRUCTURES LRFD FORMAT 9

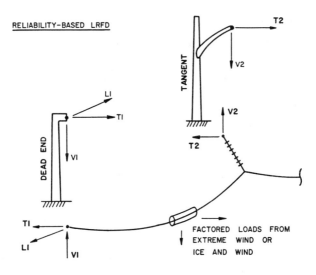

Figure 3. LRFD load factors are applied to loads on the wires

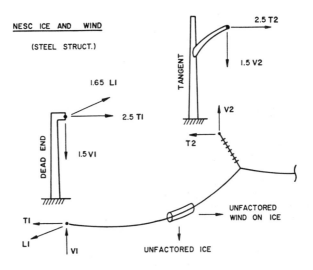

Figure 4. NESC load factors are applied to loads transferred from the wires to the structures

VII CONCLUSIONS

The reliability-based LRFD method is easy to use. Unlike the NESC, it produces reliability-consistent designs (10). It allows the designer to assign different reliabilities to different components of a line. Implementation of the method may be slowed by the current lack of hard data for certain types of components and the lack of data for ice or ice-and-wind. Once the potential economic benefits of reliability-based design are appreciated, there will be increased impetus to gather and publish statistical data on component strengths and data on ice loads. Until such time as those data are generally available, engineering judgement can be used to estimate the missing statistical properties.

ACKNOWLEDGEMENTS - The LRFD method for transmission lines is based on research sponsored by the Electric Power Research Institute and on efforts of the ASCE Committee on Electric Transmission Structures.

APPENDIX A - REFERENCES

1. Committee on Lightweight Alloys, "Guide for the Design of Aluminum Transmission Towers", Journal of the Structural Division, ASCE, Vol. 98, No. ST12, Dec. 1972, pp. 2785-2801.
2. Committee on Steel Transmission Poles, "Design of Steel Transmission Pole Structures", Journal of the Structural Division, ASCE, Vol. 100, No. ST12, Dec. 1974, pp. 2449-2518.
3. Committee on Electrical Transmission Structures, "Loadings for Electrical Transmission Structures", Journal of the Structual Division, ASCE, Vol. 108, No. ST5, May 1982, pp. 1088-1105.
4. Committee on Prestressed Concrete Poles, "Guide for Design of Prestressed Concrete Poles", Journal of the Prestressed Concrete Institute, Vol. 28, No. 3, May/June 1983.
5. Grigoriu, M, "Estimates of Extreme Winds from Short Records," Journal of Structural Engineering, ASCE, Vol. 110, No. 7, July 1984.
6. Guide for Design of Steel Transmission Towers, ASCE Manual 52, New York, NY, 1971.
7. Guidelines for Transmission Line Structural Loading, by ASCE Committee on Electrical Transmission Structures, ASCE 1984.
8. International Electrotechnical Commission (IEC), Recommendations for Overhead Lines by Technical Committee 11, 1984.
9. National Electrical Safety Code, ANSI C2, published by IEEE, New York, NY, 1977 and 1981 Editions.
10. Peyrot, A. H. and Dagher H. J., "Reliability-Based Design of Transmission Lines," Journal of Structural Engineering, ASCE, Vol. 110, No. 11, November 1984.
11. Peyrot, A. H., "Strength Data Base for LRFD of Transmission Lines," Proceedings of the ASCE Seminar on Innovations in the Design of Electrical Transmission Structures, Kansas City, MO, Aug. 1985.
12. Specifications and Dimensions for Wood Poles, ANSI 05.1, American National Standards Institute, New York, NY, 1979.
13. Guide for Reliability-Based Design of Transmission Line Structures, Research Institute of Colorado, EPRI Research Project RP-1352, 1985.

Wind Loads on Electrical
Transmission Structures

Kishor C. Mehta*, F. ASCE

Abstract

Guidelines for structural loads on electrical transmission structures were prepared by the American Society of Civil Engineers' Committee on Electrical Transmission Structures. Discussions on climatic loads that include wind, ice, and windstorms are part of these guidelines. A summary of wind loads is presented in this paper. The wind loads are discussed in terms of wind speed, probability of occurrence, height and terrain effects, gust response factors, and force coefficients. A brief discussion on tornadoes is also presented.

Introduction

Climatic design loads include effects of wind, ice, and special winds of hurricanes and tornadoes. The Committee on Electrical Transmission Structures of the American Society of Civil Engineers (ASCE) prepared a report that discusses structural loads for design of transmission lines. Discussions on climatic design loads are included in the report, which was published by the ASCE (1984). The report explains in detail components of wind loads including reference wind speed, terrain and height coefficients, gust response factors, and force coefficients. A rational design approach for ice loads and combined wind and ice is included in the report. In addition, special wind considerations such as tornadoes, hurricanes, and topographic effects are discussed in the report.

This paper focuses on wind force formula and its components. Specifically the paper presents discussions on reference wind speed, probability of occurrence of wind speed, height and terrain effects, gust response factors, and force coefficients. A brief discussion on tornadoes is included in the paper. For other loads and their effects such as ice load, galloping, and vibrations the reader is referred to the ASCE committee report (1984).

*Professor of Civil Engineering, Texas Tech University, P. O. Box 4089, Lubbock, Texas 79409.

Wind Force Formula

The wind force on the surface of transmission line components can be determined using the following formula:

$$F = A\ (0.00256 K_z V^2)\ GC_f \qquad (1)$$

where

F = force in pounds (lb)

V = reference wind speed in mph

K_z = exposure coefficient at height z above ground

G = gust response factor

C_f = force coefficient

A = solid tributary area of surfaces projected normal to wind, ft^2

The direction of force is generally in the direction of the wind. In some cases, if C_f is defined for surface area, the force is normal to the surface area.

The quantity in the parentheses in Equation 1 relates to free field wind before it interacts with the structure. The constant, 0.00256, in Equation 1 converts kinetic energy of moving air into potential energy of pressure and it also accounts for dimensional conversion of miles and hours to feet and seconds. The standardized value of 0.00256 is based on specific weight of air at 59°F at sea level.

The components gust response factor G and force coefficient C_f are related to wind-structure interaction. These two components of wind load are dependent on size, shape, and structural characteristics of the structure.

Reference Wind Speed

In the United States, the reference wind speed is the fastest-mile wind speed at 33 ft (10 m) above ground in flat and open country terrain and associated with a 50-year mean recurrence interval. The fastest-mile wind speed is defined as the average speed of one mile of air passing a wind measuring instrument (anemometer). A fastest-mile wind speed of 60 mph means that a "mile" of wind passed the anemometer in a 60-second period; a wind speed of 120 mph relates to a 30-second period. U.S. National Weather Service and most of the U.S. standards and codes use the fastest-mile wind speed.

The American National Standards Institute Standard A58.1-1982 (ANSI, 1982) gives basic wind speeds in the form of a map, as shown in

Figure 1. The map was assembled by the subcommittee on Wind Loads of the ANSI Committee A58. The subcommittee used the results of Simiu's (Simiu et al., 1979) extreme value analysis of annual fastest-mile wind speeds recorded at 129 weather stations and Batts' (Batts et al., 1980) Monte Carlo simulation of hurricane storm data. The analysis provided wind speed values that are normalized to 33 ft (10 m) above ground, flat and open terrain, and a 50-year mean recurrence interval.

There are certain regions in the country, such as mountainous terrain, where topographical characteristics may cause significant variations of wind speed over short distances. These variations of wind speed cannot be shown on a map of small scale. Special wind regions designated on the map (Figure 1) caution the designer that the wind speeds may vary significantly in these regions from those shown on the map. The designers should consult local meteorological data in these cases to establish design wind speed.

The ANSI A58.1-1982 standard permits use of regional climatic data to determine wind speed for a specific location. It is required that an acceptable extreme-value statistical analysis procedure be employed in reducing the climatic data. It has been recommended (Simiu et al., 1979) that a minimum of 10 years of data should be used to obtain dependable results from statistical analysis. Recent research by Grigoriu (1984) and by Cheng and Chiu (1985) indicates that it may be possible to determine wind speed using short records under certain climatic conditions.

Probability of Occurrence of Wind Speed

Mean recurrence interval associated with wind speed is related to probability of occurrence of wind speed. Mean recurrence interval is defined as the time, on the average, between the occurrences of a specific wind speed. Thus, the basic wind speed, which is associated with a 50-year mean recurrence interval, can be expected to occur, on the average, once every 50 years.

The probability, P, that the design wind speed will be equaled or exceeded at least once during the life of the system is given by the expression:

$$P = 1 - (1 - P_a)^n \qquad (2)$$

where

P_a = annual probability of exceeding (reciprocal of the mean recurrence interval)

n = life of the system (years)

The probability that a wind speed of a given magnitude will be exceeded increases or decreases with the period of time that the system is exposed to the wind environment and the mean recurrence interval used in the design. Values of probability of exceeding design wind

14 ELECTRICAL TRANSMISSION STRUCTURES

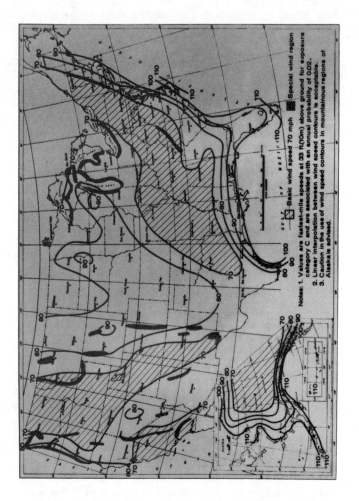

Figure 1. Basic Wind Speed in MPH (ANSI, 1982) (50-Year Mean Recurrence)

This material is reproduced with permission from American National Standard (Minimum Design Loads for Buildings and Other Structures, ANSI A58.1), copyright 1982 by the American National Standards Institute. Copies of this standard may be purchased from the American National Standards Institute at 1430 Broadway, New York, N.Y. 10018.

speed for a designated mean recurrence interval and a given design life of a system are shown in Table 1. As an example, if a design wind speed is based upon P_a = 0.02 (50-year mean recurrence interval), there exists a probability of 0.40 (40 percent) that the design wind speed will be exceeded during a 25-year life of the system. If the design wind speed is based on P_a = 0.01 (100-year mean recurrence interval), the probability of exceeding the design wind speed during a 25-year life of the system reduces to 0.22 (22 percent).

Table 1. Probability of Exceeding Design Wind Speed During Life of the System

Annual Probability P_a	Mean Recurrence Interval years	Life of the System, n (years)					
		1	5	10	25	50	100
0.04	25	0.04	0.18	0.34	0.64	0.87	0.98
0.02	50	0.02	0.10	0.18	0.40	0.64	0.87
0.01	100	0.01	0.05	0.10	0.22	0.40	0.64
0.005	200	0.005	0.02	0.05	0.10	0.22	0.40

Height and Terrain Exposure Coefficient

The exposure coefficient, K_z, in Equation 1 accounts for height and terrain effects. It is recognized that wind speed varies with height because of ground friction and that the amount of friction varies with ground roughness. The four terrain roughnesses or exposure categories considered are specified in ANSI A58.1-1982 (ANSI, 1982) as follows:

(1) Centers of large cities and very rough terrain--Exposure A.

(2) Suburban areas, towns, city outskirts, wooded areas, and rolling terrains--Exposure B.

(3) Flat, open country and grassland--Exposure C.

(4) Flat, unobstructed areas directly exposed to wind blowing over large bodies of water--Exposure D.

Detail discussions of exposure coefficient values and description of the four exposure categories are given by Mehta (1984) and in the ASCE report (1984). Values of K_z are tabulated in ANSI A58.1-1982 and are shown in graphical form in Figure 2. Values of exposure coefficient change significantly with the exposure category. For example, the values of exposure coefficient at 30 ft reduce by 50% for Exposure B and increase by 37% for Exposure D as compared to the value for

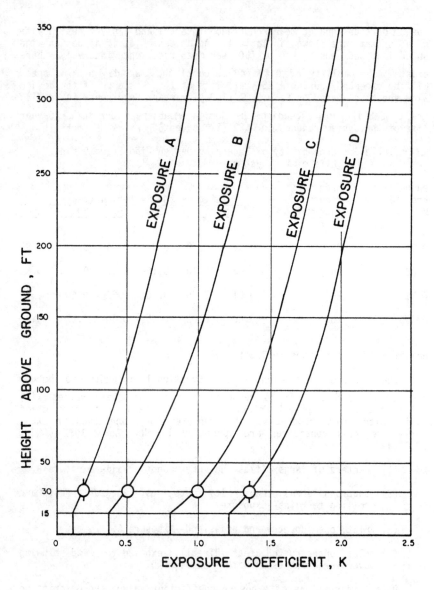

Figure 2. Exposure Coefficient K_z

Exposure C (see Figure 2). These variations in exposure coefficient value have significant effect on design wind loads since wind force is directly proportional to exposure coefficient as indicated in Equation 1.

Gust Response Factor

The gust response factor, G, in the wind force formula, Equation 1, accounts for the dynamic effects of gusts on the wind response of transmission line structures. It has been recognized that gusts do not envelop the entire span between transmission structures and that some reduction reflecting the spatial extent of gusts should be included in the gust loading. However, the resonant response of the wires and structures to wind gusts will result in dynamic amplification of the wind loadings that will tend to offset the spatial reductions. Both of these effects have been incorporated in the gust response equations developed by Davenport (1979). The gust response factors in the ASCE report (1984) are based on these equations.

The Davenport gust response factors are multipliers to the mean wind speed loading. Typical curves for gust response factors obtained from the Davenport equations for conductors in Exposure Category C are shown in Figure 3. As indicated in the figure, the gust response factors are dependent on average conductor span and the effective height of the conductor. The gust response factors are also dependent on exposure category, conductor sag to span ratio, conductor diameter, and reference wind speed. The gust response factors for structures, in addition, depend on dynamic characteristics of the structure such as natural frequency and damping. Several graphs of gust response factors for typical type, size, shape, and structural characteristics of the structure are included in the ASCE report (1984).

Force Coefficient

Force coefficient C_f is the ratio of the resulting force in the direction of the wind to the applied wind pressure. It depends on the structure's characteristics such as shape, size, orientation of the wind, solidity, shielding, and surface roughness. Because of the large number of variables and because of the inability of researchers to obtain the values theoretically, force coefficient values are obtained through wind tunnel tests and field testing. Tables and graphs of force coefficients for cables, structural members, and towers of different sizes and shapes are provided in the ASCE report (1984). These tables and graphs are used to assess force coefficient values in specific cases. The assessed values are used in the wind force Equation 1 to obtain design load.

Tornadoes

Tornadic storms can be a significant wind load factor in some parts of the country. On the average, 800 to 1000 tornadoes occur each year in the contiguous United States. These tornadoes vary significantly in their intensity. Fujita (1971) and Tecson et al.

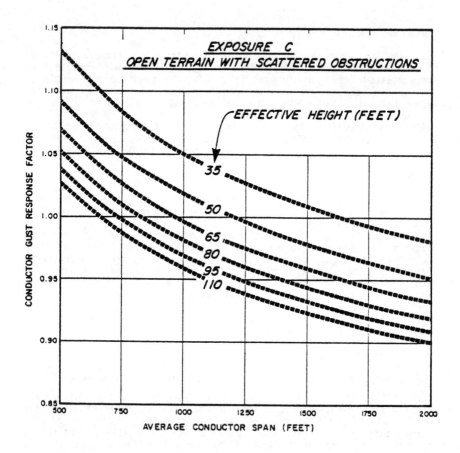

Figure 3. Typical Gust Response Factor for Conductor Based on Davenport's Equations

(1979), based on the general survey of damage, have assessed that almost 80 percent of tornadoes possess a gust wind speed of less than 150 mph. In addition, a vast majority of tornadoes cover relatively small areas. Thus, the probability of high winds due to tornadic storms for a given location is small. However, in certain areas of the country a long electrical transmission line has a high probability of being affected by tornadoes. Tornadoes may cross a transmission line and cause damage to one or two towers. This damage can be significant since it affects transmission of electrical power. Sufficient data on tornadoes and technology on tornado hazard assessment are available (McDonald, 1984) to account for tornadic storms in design.

Conclusions

There are several components in the wind force formula that require careful considerations for assessment of design wind loads. With improved data on wind speed, better knowledge of height and terrain effects, and advanced dynamic analysis for wind-structure interaction, wind loads are determined that are reliable, safe, and consistent with the economy of construction of transmission structures.

Acknowledgments

The material for weather related loads on transmission structures in the ASCE report (1984) was prepared by Working Group II under the capable Chairmanship of John W. Harrison of Black and Veatch. His leadership enabled the working group to produce a usable report. The contributions of all the members of the working group are acknowledged. The author is grateful to Dan E. Jackman of Omaha Power District for the opportunity to prepare this paper and make a presentation at the seminar.

Appendix.--References

ANSI, "Minimum Design Loads for Buildings and Other Structures," ANSI A58.1-1982, American National Standards Institute, New York, NY, 1982.

ASCE, "Guidelines for Transmission Line Structural Loading," Committee on Electrical Transmission Structures of the Committee on Analysis and Design of Structures of the Structural Division of the American Society of Civil Engineers, New York, NY, 1984.

Batts, M., Cordes, M., Russell, L., Shaver, J., and Simiu, E., "Hurricane Wind Speeds in the United States," Building Science Series Report 124, National Bureau of Standards, Washington, DC, 1980.

Cheng, E.D.H., and Chiu, A.N.L., "Extreme Winds Simulated from Short-Period Records," Journal of Structural Engineering, ASCE, Vol. 111, No. 1, January 1985, pp. 77-94.

Davenport, A. G., "Gust Response Factors for Transmission Line Loading," Proceedings, Fifth International Conference on Wind Engineering, Colorado State University, Fort Collins, July 1979.

Fujita, T. T., "Proposed Characterization of Tornadoes and Hurricanes by Area and Intensity," SMRP Research Paper 91, Department of Geophysical Sciences, The University of Chicago, Chicago, IL, 1971.

Grigoriu, Mircea, "Estimates of Extreme Winds from Short Records," Journal of Structural Engineering, ASCE, Vol. 110, No. 7, July 1984, pp. 1467-1484.

Mehta, Kishor C., "Wind Load Provisions ANSI #A58.1-1982," Journal of Structural Engineering, ASCE, Vol. 110, No. 4, April 1984, pp. 769-784.

McDonald, J. R., "A Methodology for Tornado Hazard Probability Assessment," prepared for the Division of Health, Siting and Waste Management, Office of Nuclear Regulatory Research, U.S. Nuclear Regulatory Commission, NRC FIN BB5998, Washington, DC, 1983.

Simiu, E., Changery, M.J., and Filliben, J.J., "Extreme Wind Speeds at 129 Stations in the Contiguous United States," Building Science Series Report 118, National Bureau of Standards, Washington, DC, 1979.

Tecson, J. J., Fujita, T.T., and Abbey, R.F., Jr., "Statistics of U.S. Tornadoes Based on the DAPPLE Tornado Tape," Preprints of the Eleventh Conference on Severe Local Storms, American Meteorological Society, Boston, MA, 1979.

A Field Study of
Wind-Induced Conductor Loads

By Christopher Y. Tuan[1], A. M. ASCE
Michael Thomas Potter[2], M. ASCE
Dan E. Jackman[3], M. ASCE

ABSTRACT: A 161 kV steel pole line in Omaha, Nebraska, is instrumented to provide conductor load field data due to climatic events. Various transducers and monitoring equipment used for data acquisition are described. Data obtained in field experiments are used to establish load statistics for use in reliability-based design practices.

INTRODUCTION

This study was prompted by the industry wide need for field data to support the methods and loading criteria currently in use for the design of transmission line structures. The development of reliability-based design procedures [5,9] has received recent attention in the transmission line design community due to cost effectiveness. However, the implementation of such design procedures requires probability distributions of loads and structural component strength as input parameters. The success of this implementation depends upon the applicability of the data used to establish these distributions.
Design engineers are concerned with the problems of selecting the design loads, assessing the static and dynamic loading effects, and controlling the reliability of the structure. A meaningful implementation of an economic design methodology must rely on past industry experience coupled with a working knowledge of the statistics of load and resistance data.
The National Electrical Safety Code (NESC) [8] is the leading proponent of using an Overload Capacity Factors (OLF) approach to structural design. The majority of United States utilities are required by their state law to

---

1.) Asst. Prof., Dept. of Civ. Engrg., Univ. of Nebraska-Lincoln, Omaha, Nebraska 68182-0178.
2.) Senior Engineer, Omaha Public Power District, Omaha, Nebraska 68102-2247.
3.) Manager, Transmission Engineering, Omaha Public Power District. Omaha, Nebraska 68102-2247.

follow NESC as a minimum design. This approach is rather simplistic and has major drawbacks. This has led to utilities developing their own loading agenda which incorporates local loading events that can occur in their service area.

A Load and Resistance Factor Design (LRFD) approach has been the general trend of the structural design community. Reliability-based procedures can be found in most recent structural journals. There is however a need for additional and improved loading and material data to support this recent design methodology.

This paper presents an ongoing joint research project between the Electric Power Research Institute (EPRI), the Omaha Public Power District (OPPD), and the University of Nebraska-Lincoln (UNL). The objectives of this project are: (1) to establish a data base of conductor loads, primarily due to high winds, through instrumenting an electric power transmission line in Omaha, Nebraska; and (2) based on the load data, evaluate drag coefficient, height factor, gust effect, and span effect for use in conductor load calculations; and (3) verify the adequacy of the procedure proposed in the ASCE Guidelines [6] for transmission line structural loading.

To achieve these objectives, EPRI provided a weather station with a microprocessor-based data logger, OPPD provided a 161 kV steel pole line for instrumentation and a communication network for data collection and storage, and UNL provided load transducers and technical support for data monitoring and analysis. OPPD also installed all the equipment and accessories. The duration of this joint research effort is expected to be three years.

The weather station is equipped with an anemometer, a wind direction indicator and a temperature gage. Two load cells and a swing angle indicator are installed on the insulator string attachment points of a tangent steel pole to monitor the conductor forces. These transducers convert physical quantities of interest into analog DC voltages. The analog signals are digitized simultaneously at 10 samples per second by the data logger and stored on magnetic tape. The data logger can be programmed to allow frequent scanning of data channels when a preselected wind speed is exceeded. Data will be analyzed using time series techniques [3] based on the random process theory for statistical and spectral studies.

DESCRIPTION OF TEST LINE

The test line is a 161 kV electric transmission line in a West Omaha suburban area. The line runs east-west in an open country setting having gentle rolling hills. The test pole is a 100 foot tubular steel tangent structure with three 954 MCM "Cardinal" conductors of 1.1960 inch diameter supported by steel arms with suspension insulator strings (see Fig. 1). The pole arms are on the south face

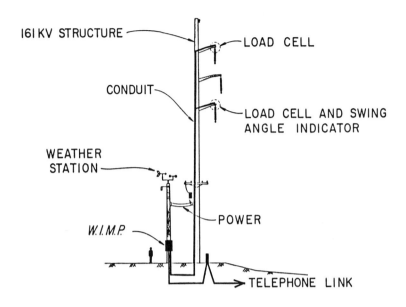

FIG. 1-TEST SETUP FOR FIELD MEASUREMENTS

of the pole. The altitude of the test site and its vicinity is about 1275 feet above sea level. The surroundings are basically corn fields with few trees or buildings to cause erratic wind patterns. Thus the terrain features of test site would fall into ANSI Exposure Category C [1]. Prevailing wind direction is from NW to SE.

Wind forces on the top and the bottom conductors are monitored by load cells installed at the insulator string attachment points. The attachment points of these conductors are eighty-five and sixty-nine feet above groundline. The horizontal span is approximately 585 feet. The lower load cell has a built-in swing angle indicator (also called inclinometer), which measures the out-of-plumb angle of the insulator string transverse to the conductor. No significant swing angle in the longitudinal direction is expected in the test line. The transverse swing angle will initially be assumed the same for both conductors monitored. With the swing angle and the force reading from the load cell, we can resolve the conductor force into vertical and horizontal components transverse to the line as illustrated in Fig. 2.

The weather station is installed on the top of a thirty-three foot steel pole located fifteen feet north and fifteen feet west of the test structure. The weather station provides information such as the reference wind speed, wind direction and ambient temperature necessary for correlation studies.

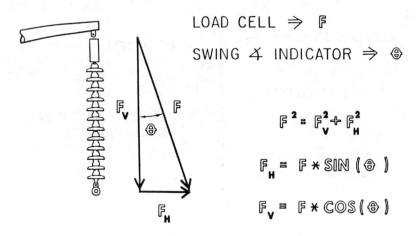

FIG. 2-RESOLVED FORCES VIA SWING ANGLE INDICATOR

DATA ACQUISITION AND COMMUNICATION

The response characteristics, accuracy and limitations of the various sensors/transducers employed in the data acquisition operation are discussed in detail in this section.

Data Logger --- The Wind Instrumentation Monitor Package, hereafter called WIMP, is a microprocessor-based data acquisition unit. The WIMP, previously known as BEASTIE (Bi-functional EPRI Atmospheric Structure Test Instrumentation Equipment), was revised and implemented by DIGI-tek, Inc., to make measurements of wind and wind-induced forces on overhead transmission lines. The WIMP accepts 6 channels of analog signals within 0 to +5 volts from various transducers to provide weather and wind force data. Signal conditioning is provided for each channel. All input signals are simultaneously sampled 10 times per second. This sampling rate is fixed and is not the rate at which data is recorded. The data recording rate is adjustable and may be as fast as once every 6 seconds. Two recording rates can be set: when the wind speed exceeds a trip value for a preset amount of time the fast rate is initiated; and when the wind drops below a lower trip value long enough, the slower rate is resumed. The trip values and time settings are adjustable. The sampling process pauses 0.1 seconds for data recording. Not all the data points sampled during an observation are recorded. Instead, only the data points at instants of maximum and minimum winds for each observation are recorded. In addition, the mean value and standard deviation of each channel during the observation are recorded. Data is recorded on a 1/4 inch magnetic tape.

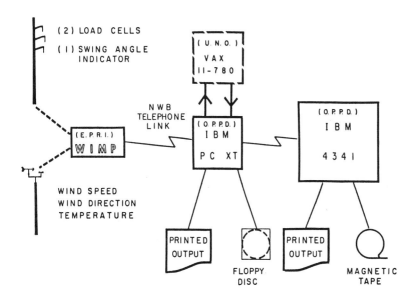

FIG. 3-DATA ACQUISITION SYSTEM

The tape has a 10 megabyte capacity. Stored data can be retrieved from the tape over phone lines via a modem hooked to an office personal computer.

Weather data sensors.--- The weather station is equipped with a J-TEC VA320 wind speed and direction sensor and a temperature gage. Instead of using the conventional rotating cups, the VA320 anemometer uses a patented ultrasonic vortex sensing technique to measure the wind speed. The sensor is mounted on a rotating vane. The vane position which indicates the wind direction is registered by a potentiometer. Wind speeds from 1 to 60 m/s (2 to 134 mph) can be measured within ±2% full-scale RMS error. The speed distance constant is 6 mm (0.24 in.). Continuous vane rotations from 0 to 360° are measured to an accuracy of ±4° at 2 m/s and ±2° above 5 m/s. The direction constant is 10 m (33 ft). The operating temperature range is -30 to 71°C (-22 to 160°F). The anemometer can survive a maximum wind speed of 100 m/s (224 mph). A thermocouple is also installed at the weather station to indicate ambient temperature for the test line.

Load transducers.--- Two Metrox 2052-10 tension load links are used to monitor conductor forces. The links have a capacity of 10 kips (2.25 kN) with ±0.25% accuracy. The links can sustain 150% overload with no damage and survive 300% overload without failure. The inclinometer has a tilt angle range of ±30° about both axes with an accuracy of less than 0.05% error. The operating temperature range is 30 to 130°F (-1 to 54°C) for the load cells and -18 to 71°C (0 to 160°F) for the inclinometer.

The transducers described above transmit data to the WIMP through six channels: two for load cells, one for swing angle, one for wind speed, one for wind direction, and one for temperature. A block diagram of the data acquisition system is shown in Fig. 3. The transducers are carefully calibrated by the suppliers before being installed. The transducers will be recalibrated every three months or as required. Analog signals from the various transducers are scanned by WIMP, which continuously stores data statistics on magnetic tape. An IBM PC XT is used to send query commands to the WIMP via a modem to retrieve information from the tape or to monitor data on-line in real time. The WIMP has a data transfer rate of 1200 BAUD. Raw data is then transmitted to an IBM 4341 and/or VAX 11/780 for detailed statistical and spectral analyses. Canned packages such as SAS [7] will be used for data reduction and analysis. Data will be analyzed at two week intervals. Raw data can also be dumped on either ASCII or EBCDIC tapes for general public use.

CORRELATION STUDIES

The primary interest of the study is to provide conductor load data due to high winds. However, it is also possible to obtain data relative to "wind on ice" situations. This possibility is discussed later in this section. The physical data acquired will be compared against analytically predicted load values using design equations. The wind force-velocity relationship proposed in the ASCE Loading Guide [6] for conductor load calculations is:

$$F = 0.00256 \ K_z \ G \ C_f \ A \ V^2 \quad \text{-------------- (1)}$$

where:

F = force in pounds,
V = reference wind speed at height z above ground
K_z = exposure coefficient at height z above ground,
G = gust response factor
C_f = force coefficient
A = solid tributary area of surfaces projected normal to wind in square feet.

Gradient wind.--- Wind speed varies with height and local terrain roughness. The exposure coefficient is applied to account for these effects on the wind velocity profile. Thus, it is equivalent to a "wind height factor". The profile of a gradient wind is normally assumed to obey a power law and the wind speed becomes constant above a certain height called the gradient height. The power-law coefficient and gradient height are specified in ANSI A58.1-1982 [4] for different topography. It should be noted that the exposure coefficient converts the pressure associated with the reference wind speed into that with the average conductor height above ground. Simultaneous force records obtained by the load cells are used to check the validity of the power-law model, while the reference wind speed is provided by the anemometer.

Dynamic response.--- Conductor forces due to turbulent wind depends upon the dynamic characteristics of the conductor with its supporting structure as well as the peak magnitude and spatial scale of the gusts. It is well known that the size of gusts is usually small compared with wind spans of conductors. Thus, the spatial distribution of gusts is an important design consideration. Two approaches are generally followed to estimate gust response of a conductor. One approach is to apply a "gust factor" to the mean wind speed to find the peak gust, a "span reduction factor" is then applied to adjust the conductor force calculated based on the peak gust speed. A second is to apply a "gust response factor" directly to the conductor force calculated from wind pressure to account for additional effects due to wind turbulence and dynamic amplification at a natural frequency of the conductor.

Gust factor is the ratio of peak gust speed to the mean wind speed. Apparently its value depends on the averaging periods for the gust and the mean wind. However, no standard averaging period has been proposed for wind measurements. Since during each observation wind speed is digitized with time interval of 0.1 second between data points, gust factors based on a number of combinations of averaging periods, e.g., 2-sec gust on 10-min. mean wind, can be established.

Span reduction factor is a function of terrain roughness, effective conductor height, wind speed, and span length. Based on the conductor forces measured, span reduction factors can be computed for prescribed gust factors. Analytical expressions for gust response factors have been derived by Davenport [4] based on the random process theory. His gust response model takes into account the power spectral densities and spatial correlation of wind and the dynamic admittance of the transmission line system.

Drag force.--- The pressure differential between its windward and the leeward surfaces results in a drag force on a conductor. This drag force is related to conductor's aerodynamic behavior which depends on the shape, size, surface roughness of the conductor, and the angle of incidence of wind. Drag coefficients are often determined from wind tunnel tests which are used to convert wind pressure into effective force on conductors. Sakakibara et al. [10] have reported that drag forces can be minimized by optimizing conductor's surface texture. Reduction of wind pressure up to 95% has been achieved for conventional size conductors in field experiments under natural wind conditions.

Oblique wind.--- When the angle of incidence of wind is not perpendicular to the conductor, the wind is yawed with respect to the line. Data provided by the wind direction indicator are used to define the angle of yaw Y. The reference wind speed is then multiplied by cos(Y) to find the component normal to the conductor for use in the wind force calculations.

Wire tension.--- The tension induced in a flexible cable by its own weight and by specific loading conditions can be calculated using the catenary formulas [2] and/or readily available computer programs. The horizontal component of the tension is constant throughout the cable because the resultant tension must act in a direction tangent to the cable. Thus, the tension is maximum at one of the two attachment points where the slope is steepest and minimum at the lowest point of the cable.

Wind force on ice.--- During icing season (late October thru early April), ice loading on conductors due to glaze accretion may be significant. The thermocouple will provide good indication of ice formation on conductors. Glaze is usually formed from freezing rain with temperature in the range of 24° to 32°F. When ice accretion on conductors is confirmed, the conductor loads provided by the load cells may be correlated with wind speeds to evaluate "wind on ice" effect on the conductors.

CONCLUSIONS

The structural engineering community has been a primary user of weather data for many years. Climatic information is available from various sources such as the National Climatic Center, Nuclear Regulatory Commission, Environmental Protection Agency, etc. These are generally not collected in a format which readily satisfies engineering design needs. It is important to promote the acquisition of climatic load data to provide quality information for design engineers.

A data base of conductor loads relative to Type C exposure will be established through this field study. Load statistics and spectra will be presented for use in the design of transmission line structures. Data bases associated with other types of exposure in other areas are also needed for correlation studies with Type C data. This is necessary for the advancement of reliability-based design methodology.

REFERENCES

1. ANSI A58.1-1982, "Building Code Requirements for Minimum Design Loads in Building and Other Structures," American National Standards Institute, New York, N. Y., 1982.
2. Beer, F. P., and Johnston, E. R., Vector Mechanics for Engineers: Statics, McGraw-Hill, Inc., New York, N. Y., 1972, pp.282-284.
3. Bendat, J. S., and Piersol, A. G., Random Data: Analysis and Measurement Procedures, John Wiley & Sons, New York, N. Y., 1971.
4. Davenport, A. G., "Gust Response Factors for Transmission Line Loading," Proceedings, the Fifth International Conference on Wind Engineering, Colorado State University, Fort Collins, July 1979.
5. Goodman, J. R., Vanderbilt, M. D., and Criswell, M. E., "Reliability-Based Design of Wood Transmission Structures," Journal of Structural Engineering, ASCE, Vol.109, No.3, March 1983, pp.690-704.
6. Guidelines for Transmission Line Structural Loadings, Committee on Electrical Transmission Structures, ASCE, the Structural Division, 1984.
7. Lewis, B. R., and Ford, R. K., Basic Statistics Using SAS, (SAS-Statistical Analysis System, developed by SAS Institute), West Publishing Company, St. Paul, Minnesota, 1983.
8. "National Electrical Safety Code," ANSI C2, IEEE, New York, N.Y., 1977, 1981 and 1984, Eds.
9. Peyrot, A. H., and Dagher, H. J., "Reliability-Based Design of Transmission Lines," Journal of Structural Engineering, ASCE, Vol.110, No.11, November 1984, pp.2758-2777.
10. Sakakibara, A. et al., "Development of Low-Wind-Pressure Conductors for Compact Overhead Transmission Line," IEEE Transaction on Power Apparatus and Systems, Vol.PAS-103, No.10, October 1984, pp.3117-3124.

"STEEL POLES - PAST, PRESENT AND FUTURE"

Ronald E. Randle*
M.ASCE

Today's tubular steel pole industry was born approximately tweny-five years ago. Since its birth, evolution has brought about numerous changes in the design and manufacture of tubular steel poles.

While the use of tubular steel poles for electrical transmission line construction is a newer concept than either steel lattice or wood, it is in its own right a mature industry. Evidence of its maturity can be found in it sophistication. Sophistication in both design and manufacturing production is the trademark of today's successful companies.

Sophistication has been borne out of innovation. It was innovation that gave the industry its beginning, it has been innovation that has allowed the industry to grow and prosper and it will be innovation that will provide the industry its momentum for continued expansion in the future.

DESIGN:

During the early years, steel pole design was little more than a crude and conservative application of the most basic of structural analysis principles. These early designs were done manually using a trial-and-error approach. The mid 1960's saw the first generation of computer software which would design a simple cantilever structure using linear analysis methods. In the years following, very significant enhancements were made to the computer analysis methods of design where today, the software is extremely sophisticated.

Internally developed programs are being used by most all the pole fabricators to design the vast array of tubular steel structures required by the utility industry. Non-linear analysis techniques are state-of-the-art. And, optimization is no longer being done solely on the basis of structure weight but rather on cost of structure fabrication.

*Vice President RD&E, Meyer Industries,
 P.O. Box 114, Red Wing, MN 55066

Because each fabricating plant is layed out differently from all the others with unique equipment combinations and thus unique capabilities and limitations, designs for any given structure are almost always different from one vendor to the next. Each fabricator has developed design details and manufacturing procedures independent of the others. Some standardization within the industry is possible, but complete standardization is not practical. For this reason, the industry practice of having each vendor submit "custom designs" for specific requirements is, and will continue to be, the basis for the majority of steel pole procurements.

During the course of these past twenty-five years, several changes have taken place in the fabricating process. And, as much as anything else, it is these changes which have affected the changes in our design process.

MANUFACTURE:

The earliest manufactured poles were often made by butt welding several short length tubes together. A fabricator might have used a dozen or more tube sections welded end to end to make up a one hundred foot pole. In time, long presses were purchased which enabled the fabricator to minimize the number of sections needed for a given pole. Also, tandem press arrangements were set up which further reduced the number of tubes needed for pole production. Today, the same one hundred foot pole might be fabricated using only two tubes and a single circumferential weld. (See Figure 1.)

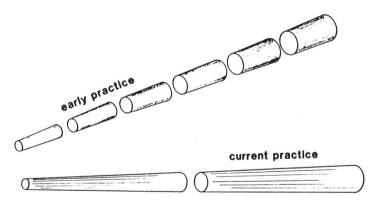

Figure 1. Steel Pole Design - Past & Present

In order for a manufacturer to be able to fabricate these longer tube sections, it became necessary for him to modify existing equipment or to acquire new equipment. In the early years of pole production, plate for tapered tube sections was often cut into trapezoidal shapes using either a shearing operation or an oxy-fuel flame cutting operation. Technological advancements in the area of computer-aided-manufacturing has changed these operations dramatically. Today, numerically controlled machines direct multi-headed torches in the cutting of tower plate, baseplates as well as many of the hardware shapes. The accuracy provided by this equipment has solved almost all the problems associated with fitup and also provides much improved weld joint profiles. Along with the new equipment used for plate cutting, new welding equipment was also installed for making the longitudinal seamwelds. These semi-automated welders significantly reduced the actual weld time as well as improving the weld quality.

Improvements in the manufacture of steel poles have been many. What was once a labor intensive product, is no longer; the time spent to fabricate poles today is significantly less than what it was twenty-five years ago.

THE FUTURE:

During these first twenty-five years, there have been many changes in both the design and manufacture of tubular steel structures. Still, there are many changes to come. Some will be highly visible, while others will be all but invisible.

Changes in design philosophy often prove to be the invisible types of changes. The merits of one such proposed philosophy change are currently being debated. This change would allow steel design to extend beyond the elastic limits of the material. The elastic theory of analysis has been used for design almost exclusively throughout the history of the steel pole industry. However, certain advantages could be realized if an elastic-plastic theory were accepted for certain accidental and/or emergency loading situations. These are loads which are not expected to occur but which cannot be ignored either. Examples can be found where these loads controlled the design of certain structure components, or even the entire structure. In these situations, the structures were really overdesigned for the normal and extreme loading requirements which were expected to occur.

Should this practice continue, or can we redefine failure and accept plastic yielding or even buckling under certain conditions? Input from all areas of concern is needed to establish acceptable design criteria. Also, research needs to be done to establish strength limits of the materials and member shapes in question.

STEEL POLES

The most visible types of changes are those that involve structure configuration. With welded construction, almost anything is possible. Already, in this industry's short history, we have seen numerous configurations of structures. Some of the more interesting innovations in structure configuration of late have involved hybrid construction.

Commonwealth Edison has utilized the hybrid concept of construction to achieve some specific objectives. The particulars of the configuration which make their structure interesting and/or unique are shown in Figure 2.

These particular structures are some of the largest single pole structures to be installed using direct embedment techniques with conventional granular backfill material. The majority of these embedded structures have design ground line moments of 3000 to 3500 ft-kips. And, there are a few structures which have groundline moments of between 4000 and 6000 ft-kips. The structures themselves have been designed to withstand the effects of 3 to 3-1/2 degrees of foundation rotation under critical loading situations.

The lattice arms were another unique feature about this particular structure. Earlier structure designs for Commonwealth Edison utilized braced tubular arms. Because of their lengths, these arms made access difficult without special provisions for the linemen (e.g. catwalks). This is what prompted Commonwealth Edison to consider substituting lattice type arms for the tubular arms. Their operating personnel were familiar and comfortable with the

345kv − single pole
double circuit structures
lattice arms
long span construction
direct embedded
design assumes fndn rotation

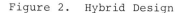

Figure 2. Hybrid Design

features provided by lattice arm construction. And, the cost of the lattice arm construction was lower than for the tubular arms with catwalks.

During the design phase, structure designs were proven out both analytically and through a series of full scale load tests. When the structure was tested however, the pole was not direct embedded. It was determined that the soil conditions at the test site were not the same as the conditions where the line would be installed and even if the soils were similar, a location could not have been found where the soil was undisturbed. Thus, direct embedment of this test structure was deemed inappropriate.

Another type of hybrid construction which is a bit more futuristic involves the combining of steel components and fiberglass reinforced plastic (FRP) components. For example, a structure could be created using a single tubular steel shaft with davit style FRP arms. One advantage afforded by this type of construction might come from the insular value of the FRP. Whereas, steel is a conductive material requiring long strings of insulation to separate the conductors from the steel, FRP is basically a non-conductive material. This could yield significant savings in the cost of insulation required.

The state-of-the-art in FRP production today does not yet lend itself to direct conversion from steel components to FRP components. However this is not because of strength. FRP product is available with flexural strengths in excess of 60,000 psi. Production techniques are the major problems. Today, FRP products are only cost effective when purchased in large quantities per item. Certain FRP items are available to our industry today. However, these are catalogue items which are mass produced. In time, production techniques will be refined to the point where some customization of FRP products will be both possible and cost effective.

Within the steel industry itself, the emphasis on R&D remains very high. Every steel mill is constantly seeking to improve their present product lines and to develop products for new markets.

The steel pole industry has and should continue to be helped by these efforts. The steel makers have placed much emphasis on cost reduction while at the same time striving to maintain or improve the product's quality. Today, there is a revolutionary new product which could dramatically impact our industry by effecting wholesale changes in both the design and manufacturing processes. This new product, which was born offshore, can effectively be described as "tapered plate". The tapering of the plate does not describe changes to the width or length dimensions, but rather the thickness.

From the very beginning, tubular poles have been made cost effective by custom designing them using changes in plate thickness along the pole's length to fit the loading requirements. A material savings is realized but labor has to be added to join the two tube sections together. Now, with tapered plate, the same type of material savings can be realized without the addition of this labor.

The tapering configurations possible off these specialized mills are almost limitless. Figure 3 shows some of the typical profiles which might be utilized in the design and manufacture of tubular steel poles.

This technology is new and product is not yet available from any of the domestic mills. However, the product is available from offshore sources at competitive prices. It is expected that technology transfer is only a matter of time and tapered plate may soon thereafter be a major feature in the production of tubular steel poles.

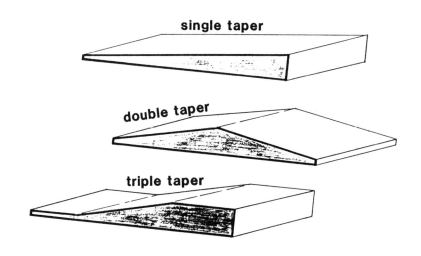

Figure 3. Tapered Plate

SUMMARY:

When using steel pole structures, construction possibilities are limitless. Of all the options available, tubular steel structures provide the most versatility for transmission line construction. Two main factors account for the truth in this statement: First, within the industry, the manufacturers work hard to capitalize on the new and innovative concepts being marketed by its raw material suppliers. Second, the manufacturers have and continue to invest capital in new equipment and processes to keep their manufacturing plants modern and efficient. Still, a third factor is essential to bring the goal of line design optimization within reach. This ingredient is open-minded utility participation.

Utility personnel, in particular the engineering staff, must be willing to review and honestly weigh the merits of new concepts in structure and line design. Relying solely on old standards for new line design is a sure way to discourage innovation. And, without innovation an industry will not survive for very long. The U.S. utility industry can ill-afford to lose a segment of suppliers for this reason. Thus, it is in everyone's best interest for utilities and its suppliers to work together. Both factions need to be receptive to new ideas.

LATTICE TOWER DESIGN USING COLD-FORMED SHAPES

by Paolo Faggiano *

Abstract

The use of cold-formed shapes in lieu of hot-rolled angles may lead to major changes in the design of lattice transmission towers.The design procedure of the new shapes is more creative, more complex but computer based, thus modifying the role of the designers.The tower outline is affected by the peculiar characteristics of cold -formed shapes and by the importance of connection detailing, giving rise to a wide range of structural solutions which allow to meet specific requirements of each situation. The first part of the paper deals with some aspects of this new design philosophy.
In the second part three well known tower configurations, i.e. guyed towers rigid towers and portals, are taken into consideration. For each configuration a solution adopting cold formed shapes is presented along with the discussion of the specific requirements on which the design is based.

Introduction

Latticed towers for transmission lines have been traditionally built up with hot-rolled steel angles. During their history, steel latticed towers have substantially changed in dimensions, shape and structural upsets. In addition steel quality has been improved together with the protection against corrosion. But for many decades the hot-rolled angle has been, by far, the most used structural element. This allowed to collect a unique deal of theoretical, experimental and in field experiences on the behaviour of such members and structures. As a consequence, at present, good tower designers are perfectly aware of the performances and limits of angles, being able to control the numerous variables involved in the design, and to choose in a short time the minimum cost structural solution in any situation.
The use of cold-formed shapes in lieu of hot-rolled angles may lead to major changes in the design of latticed towers. A wide range of different shapes, which should be designed with new procedures and shows different characteristics and specific requirements makes the difference with respect to hot-rolled angle towers. But the use of these shapes also opens new prospects as regards latticed towers.

* Research Engineer, SAE S.p.A., Via Fara 26 I-20124 Milano, Italy

On the basis of a six year experience in this field, in the following the major changes in the design will be discussed and non-conventional structural solutions made actual by cold-formed shapes for three tower configurations will be presented.

Design procedures

Theoretically speaking, cold-formed sections can be fabricated in a numberless series of different shapes and dimensions. For practical use the field is greatly restricted. Present practice for transmission towers indicates a minimum thickness of 3 mm., whereas economical considerations lead to an upper limit of thickness of about 8 mm. Further restrictions to the range of shapes are due to fabrication and detailing considerations.

Nevertheless the variety of shapes and dimensions provided by the cold- forming process and which are suitable for application on towers is much wider than the present list of hot-rolled angles.

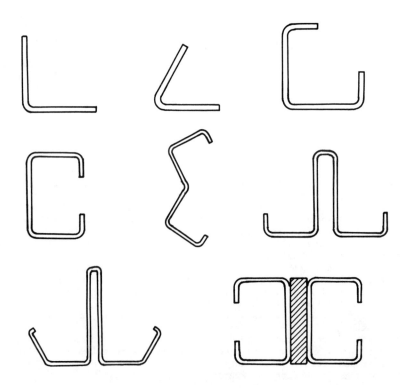

FIG.1 - Cold-Formed Shapes Used in Transmission Tower Design

In order to simplify the following design steps, it may be useful to create a reference list of sections (shapes and proportions) meeting the above mentioned requirements and which are selected according to their efficiency in various load carrying capacity versus unsupported length fields.

Figure 1 shows a number of shapes which have been used in transmission tower design.

The design path for the determination of the load carrying capacity of cold-formed members differs from that used for angles [4][5]. In checking against local buckling it must be taken into account that these shapes are composed of both stiffened and unstiffened elements which should be dealt with in different ways. Furthermore the overall buckling mode is not necessarily flexural but can be a combination of flexure and torsion or simply torsional, thus involving specific calculations. Consequently nearly every cold formed shapes has its own instability curve. Therefore, the more cumbersome design path to be followed for cold formed members implies that it should be computer-based. The role of the tower designer then changes.

At present, the selection of the sections can be made manually on the base of an elastic analysis and of a limited series of instability curves of general validity. Detailing is based on a wide number of standard connections and draftman consults the designer only for special problems.

With cold-formed members the calculation of the load carrying capacity of the single shape is made automatic and is out of control of the designer who should take care of other important aspects.

Mainly he should select the shape which is suitable for application in the various locations of the structure. In doing this be should take into account that if there are different shapes with similar efficiency, the choice of the least complex leads to simplifications in fabrication. But, above all, he should take into account the possibility of connections of the shapes which meet at one point. This requires a tighther interaction bewteen the designer and the draftman since the most efficient solution is obtained when design and detailing are developed at the same time.

Cold-formed shape characteristics

The main advantage of cold-formed sections is that they can be shaped and proportioned so as to fit the effective working conditions of each member in the tower. According to the various member locations, shape and dimensions can be choosen in order to improve the local buckling strength, the flexural buckling strength and the torsional and warping strength.

From the experience, it emerges that in general cold-formed shapes can be efficiently used with unsupported lengths higher than those used for equivalent hot-rolled angles. This results in a reduction of the tower weight, in a thinning out of the scheme and in a reduction of the number of connections. The importance of this peculiarity should be emphasized. First of all cold-formed member towers are lighter and consist of a lower number of bars, this leading to cost savings in fabrication, handling and erection.

But, connection detailing with cold-formed members is generally more complicated. With thin members, the bearing on member holes often controls the design and the number of required bolts per connection tends to increase. In addition, in order to realize concentric connections of complex shapes a higher number of gusset plates and bolts may be necessary. Nevertheless, the above mentioned reduction of the number of connections generally leads to an overall reduction of plates and bolts, this obviously being less sensitive than the reduction of the number of bars.

Another remarkable characteristics of cold-formed shapes can be referred as flexibility. Figure 2 shows a tower weight versus number of member plot in which points representing various design alternatives for both hot-rolled angle towers and cold-formed member towers are reported. It can be noted that the curve referring to cold-formed members lies below the one referring to hot-rolled angles and increses at a lower

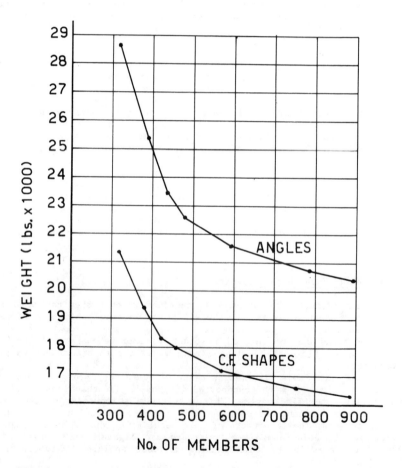

FIG. 2 - Design Alternatives with Angles and Cold-Formed Shapes

rate for decreasing number of members. Looking now at the overall cost of a tower, this is composed of various items (design, fabrication, transport, erection) which are affected in different ways by design parameters (weight, number of pieces) depending on each environmental situation.

From the above mentioned curves it can be concluded that with cold-formed shapes there exists a wider range in which the optimum combination of the design parameters can be found and that the resulting cost gap can be used to meet specific requirements of each design situation.

Tower configurations

So far, major changes in the design caused by the adoption of cold-formed members have been briefly reviewed.

In the following three interesting applications of these shapes to different tower configurations are presented, along with the discussion of the specific requirements on which each design was carried out.

FIG. 3 - V Guyed Tower for EPRI Experimental Line

Guyed tower

It is universally acknowledged that guyed towers are an advantageous solution especially for UHV lines, i.e. when the weight and dimensions of rigid towers become considerable, on flat terrains. The major advantages of this solution are the low weight, the savings in foundation cost and the possibility of adopting very efficient erection techniques by light cranes or helicopters.

The main members of a guyed tower mast are subject to different axial loads due to the combined action of axial forces and transversal (wind on mast) forces. For instance, in a guyed V tower inward post members are less loaded than outward ones, and the axial load on them can be carried by a single member. Thus the adoption of a triangular cross section with the single leg member inwards is an efficient solution. The advantage is that a complete face is eliminated thus leading to a lower weight and a lower number of pieces of the tower.

This solution was adopted in a prototype for the 500 kV experimental line of EPRI which will be erected at Haslet within the end of this year (fig. 3).

The V type guyed tower was designed to hang conductors in a triangular "inverted delta" configuration which allows for very good electrical performances. This tower differs from a conventional CRS tower, for the horizontal strut which eliminates the effect of the vertical loads on the guys and on their foundations [3].

Both the masts and the strut have an equilateral triangular cross section. The main members of the masts have a "W" shaped section with stiffeners with outer legs forming a 60° angle. A staggered bracing system was adopted for the masts. The behaviour of the main member was analyzed by a specific computer program [2] which calculated the critical axial load by an energy method, taking into account the stiffness of the bracings. The full scale test of the tower confirmed the validity of the approach.

Special care was devoted to the design of the main connections, like leg member splices and bottom and top connections of the masts, in order to reproduce the design working conditions. The comparison between this tower and the equivalent hot-rolled angle tower shows a 10 per cent weight reduction and 46 per cent less members.

Self-supporting tower

The full cost of a transmission line consists of various items. The cost of foundations is one of these items which has a big and increasing incidence especially for rigid towers of EHV and UHV lines.

The reduction of foundation costs was the goal for the design of a rigid tower for a 380 kV line of ENEL, the Italian Power Authority. Some prototypes of this tower were used, together with standard ENEL towers, in a line in central Italy (fig. 4).

To achieve the goal the tower body was designed with an equilateral triangular cross section and, therefore, with three legs instead of four. Since, with a triangular base, only two foundations are stressed by moments due to transversal loads (which usually control the design) the width of the base was increased with respect to the standard tower in order to have similar loads in the foundations. The third footing was then less stressed since it has to balance only longitudinal moments.

FIG. 4 - ENEL 380 kV Triangular Suspension Tower

Table 1 - Comparison of Foundation Costs

	FOUNDATIONS					
	CHARACTERISTICS				COSTS	
	No.	EXCAVATION ft.^{3}	CONCRETE ft.^{3}	STEEL lbs.	TOTAL $	%
STANDARD TOWER	4	1860	472	1298	4337	100
TRIANGULAR TOWER	2+1	1413	321	1128	3147	72.6

Note: 1 ft.^{3} = 0.028 m^3; 1 lb. = 0.453 kg

The leg members were realized with bent "T" shapes, with outward flanges forming a 60° angle.

It should be noted that the upper portion of the tower has a square section due to geometrical problems for both connections and stiffening efficiency. From the point of view of detailing, then, the part between the body and the upper portion, in which the cross section changes from triangular to square, was the most critical. This because there is a number of joints, in which several members from different directions converge.

The foundation cost comparison with the standard tower (table 1) for dry soils shows a reduction of about 27%. It was performed on the basis of ENEL standard foundations, and does not take into account the extra savings of one less positioning of drilling machines.

It should be emphasized that, further to these sensible savings, the weight of the tower was also slightly reduced (a few percent) and the number of members was 29% lower. The line was erected and strung last spring and is now in service.

Portal tower

It is well known that latticed towers are a cost effective solution for long span lines, with fewer towers, insulators and fittings, and where a sensible longitudinal strength is required. The field of short span transmission lines , ranging approximately from 115 kV to 345 kV, which are widely used in USA, is dominated by wooden or steel tubular poles or "H" frames. Lattice tower designers have been looking for a long time, for a competitive solution from both the economical and aestetic point of view, but results have not been satisfactory.

So far, even if the weight and, in most cases, the supply cost of lattice towers is the lowest, assembling and erection costs are sensibly higher and the cost comparison is overturned, resulting favourable to

FIG. 5 - Complete Set of Elements for a 230 kV "H" Frame

wooden or steel tubular structures.

The problem was faced when designing an "H" frame tower for a 230 kV line of Florida Power Corporation [1]. The first design constraint was relative to foundations. Masts had to be directly embedded in a drilled hole in the ground, and mast dimensions had to fit a 3' diameter hole, this being the diameter of the available drill. In order to cut assembling costs on site it was decided to pre-assemble elements in the workshop and to ship them by containers to safeguard their integrity. This was an efficient solution since pre-assembling operations in the workshop can be rationalized and cost much less than on site. The shape of the cross section of the mast was then designed taking into account the maximization of the number of elements stowed in a standard container.

The best solution for both strength and transport resulted a right angle isosceles triangle with special "W" shaped elements at the 45° corners and a plain angle at the 90° corner. Furthermore, for transport purposes the cross-arms and the bridge were studied to be folded by

FIG. 6 - 230 kV Line with Latticed "H" Frames

loosening some bolts. Figure 5 shows the elements which constitute a complete tower whereas figure 6 shows the erected tower. The design led to a cost-effective result, and the towers due to their light weight (about 25% lower than wooden structure), were very efficiently erected by helicopter.

Conclusions

The use of cold-formed shapes in the design of transmission latticed towers implies some changes in the designer approach, due to new design procedure and to the different characteristics, in the research of the most efficient application of such shapes.
On the other hand this technology allows for a wide range of new non- conventional solutions which improve the cost-effectiveness of lattice towers also in those fields where, traditionally, this type of structure is not competitive.

Appendix I - References

1. Agostoni P, Casarico G. "Una alternativa in profili piegati a freddo alle strutture in legno per il trasporto di energia a 230 kV negli Stati Uniti: realizzazione e prospettive", Proceedings X° convegno C.T.A., October 1985.

2. Catenacci A., "Stability of thin walled bar with intermediate elastic support (theoretical formulation and numerical investigation)", Costruzioni Metalliche N° 2, 1985.

3. Catenacci A., Faggiano P., "Nuove tipologie di sostegni in profili piegati a freddo per linee elettriche a 500 kV: la linea sperimentale dell'Ente di Ricerca Elettrica Americano EPRI", Proceedings X° Convegno C.T.A., October 1985.

4. Gaylord E.H., Wilhoite G.M., "Transmission Towers: Design of Cold-Formed Angles", Journal of Structural Engineering, Vol. 111, N°8, August 1985.

5. Zavelani A., Faggiano P., "Design of Cold-Formed Latticed Transmission Towers", accepted for publication on Journal of Structural Engineering.

INNOVATIONS IN PRESTRESSED CONCRETE STRUCTURES

William M. Howard,[1] M. ASCE

ABSTRACT

Prestressed concrete poles are probably the best choice for most transmission line structures. They have many advantages over structures made from other materials for both engineering and economic reasons.

One of the primary engineering benefits of prestressed concrete transmission line structures is the tremendous resistance which these structures provide against cascading failures. The sources of these benefits are:

1. The inherent flexibility of prestressed concrete poles.

2. The ability of concrete to withstand large overloads for the short duration of shock loads.

3. The large mass of the pole which tends to resist the peak shock forces lasting only a fraction of a second.

4. The protection of adjacent structures due to the partial longitudinal restraint provided by even a broken pole because the steel in the pole holds the broken parts together.

In order to account for these advantageous features in the design of a line, it is necessary to consider the structures and the conductors as a system and to analyze the system by means of an advanced computerized analysis such as SAPS (3) or BRODI2/BROFLX (1). Although this type of analysis is not commonly done, it is probably even more important to the integrity of the line than the normal analyses.

Just as a complete engineering analysis is necessary to realize the engineering benefits of prestressed concrete poles, a complete economic analysis is required to realize the economic benefits. Life Cycle Costing and Present Value Analyses must be utilized to insure selection of the best structures and avoid economic failures which, although not as spectacular as structural failures, are as damaging to the utility and the consumers.

----

1. President, Power Line Systems, Inc., 6701 Seybold Road, Madison, WI 53719

ENGINEERING CONSIDERATIONS

DESIGN PHILOSOPHY

A fundamental precept of this paper is that power lines will fail. It is not possible to make them absolutely fail-safe and even if it were possible, it would be economically unwise. Therefore, the wisest course is to design lines which provide an acceptable degree of reliability in everyday use and then to insure that when a failure does occur, it is contained so as to avoid massive failures. The new Load and Resistance Factor Design (LRFD) (2) techniques refer to the concepts of Safety, Reliability and Security with the term Security meaning the ability of a line to contain failure damage in such a fashion that there is little or no damage to surrounding structures when any given structure fails.

When we design a structure, we usually look at such loads as NESC, Extreme Wind, Heavy Ice and Broken Conductor. NESC loads are included in the definition of Safety, since these are the minimum conditions that the code allows for "safe" operations. The climatic loads such as Extreme Wind and Heavy Ice would come under the definition of Reliability. If we don't design adequately for these loads we are likely to have an unacceptable level of failures. Finally, in an attempt to provide security for the line, we look at broken conductor conditions in which the broken conductor longitudinal load is frequently equated to the maximum conductor tension. The problem is that rather than treating the structure as part of a structural system, we usually apply the broken conductor load to an isolated structure and design the structure so that it will support the load. This results in a structure that is grossly over-designed in comparison with our expectations.

To a large degree, all structures are guyed structures wherein the conductors act as guys in a longitudinal direction. It is this fact that is usually forgotten or ignored. As long as most of the conductors remain intact and secure at the dead-end points and the structure has at least a reasonable degree of longitudinal strength, it is unlikely that a catastrophic failure will occur. The most serious types of failures are not the initial structure failures caused by high wind, heavy ice, a broken conductor or other trauma. They are due to the sudden loss of tension on all of the conductors which can occur, for example, when a heavy angle or dead-end structure fails.

Considering the importance of a longitudinal analysis, it is amazing how seldom it is done. It not only should be done, it is relatively easy to do it or to get it done. For those with access to an EPRI work station or a main frame computer, there is the BRODI2/BROFLX program. The SAPS program is a more powerful program with graphics output which can be run on an IBM PC. Finally, for a reasonable cost, a consultant can do an analysis.

The next important factor to examine is how a line built of prestressed concrete poles works to contain failures and keep them from becoming cascading failures.

FLEXIBILITY - THE KEY FACTOR

With sufficient thought, most utility engineers are aware of how rapidly sag and tension change with a change in the length of the conductor between two support points. In the course of sagging wire, for example, the final adjustments are measured in "clicks" on a chain hoist. An inch of difference will make a substantial difference in sag and tension. Consider, then, what happens if a corner structure gives way in a line constructed with very flexible prestressed concrete structures.

By essentially removing the conductors from one side of a tangent structure, it now tries to act like a dead-end structure. The first thing that happens is that the insulators move. In the case of a suspension string, it swings. Fiberglass post insulators will bend around the pole and may be severely damaged but they generally do not separate. Porcelain posts may break off, in which case the structures will not cascade, but the insulators and conductor may. This is the big shortcoming with rigid, porcelain post insulators without properly working load limiters.

Once the insulators swing, a significant reduction takes place in the conductor tensions. However, there is still a large unbalanced force to work on the structure. If the structure is rigid, as in the case of lattice steel towers, either the tangent structures must be built strong enough to withstand this residual tension (plus an impact force) or there will quite likely be a cascading failure. Either proposition is very costly. Fortunately, in the case of flexible prestressed concrete poles, a large deflection is possible before the pole fails. This deflection, even more than the relatively short insulator swing, will greatly relieve the residual tension.

An example in Appendix A shows how these principles work so that one can see the significance of the numbers.

SHOCK LOADS

In addition to the residual loads discussed in the previous paragraph, there are shock loads that must usually be dealt with any time the conductor tensions are released. In the case of prestressed concrete poles, there are two factors that work to mitigate the problems associated with these shock loads.

In the first place, these shock loads are very short term (lasting only a fraction of a second). The very large mass of these poles means that tremendous inertial forces are mobilized if the top of the pole is moved a distance of several feet (typically 10% of the pole length) during the short duration of the shock load. In essence, the pole itself is not affected by the shock load. It feels only the difference between the shock load and the inertial force of the pole.

Even if some of the shock load should be applied to the structure, a prestressed concrete pole has the ability to absorb such a load due to the nature of concrete failure. Since concrete fails through a micro-cracking process which is time dependent, it has the ability to withstand up to a 20% overload for the short duration that these shock loads exist.

THE FINAL PROTECTION

Even if a corner structure fails and the adjacent tangent structure is not quite able to hold the resulting load, when it breaks, the top portion cannot be separated from the portion that is buried in the ground. Unlike wood poles which snap off and travel along the right-of-way with the conductors, the concrete poles simply fall down and act as anchors attached to the conductors. These conductors then act as a partial restraint or back-guy for the adjacent structure so that it does not see the full load that broke the previous structure.

ECONOMIC CONSIDERATIONS

LIFE CYCLE COSTING

Although the total lifetime cost of a line built with concrete poles is almost always less than for lines built with other materials, this advantage is frequently not understood. Few utilities do any life cycle costing or, for that matter, have sufficient background data to conduct proper cost analyses. The primary costing concept in the industry seems to be one of a simple first cost of materials comparison.

In order to make a proper economic analysis, there are many things to be considered other than first cost. Different product lives and varying maintenance costs are important. This means that accurate historical data needs to be available. One needs to know the life of paint on steel poles and the cost to repaint them. It is important to know the average lives of wood poles and the cost of maintenance required to achieve those lives (e.g. the cost of inspections and treatments). Another major cost to be determined is the cost of tightening hardware made necessary by the shrinkage of the wood. Finally, the cost to change out a deteriorated wood pole will impact heavily on the life cycle cost. For example, a person might not pay much of a premium for the longer life of a concrete pole if the wood pole is in a tangent location with easy access and can be changed out with a four man crew in two or three hours. However one can afford to pay a very large premium for a concrete pole if it is to be part of a dead-end structure in the middle of a swamp requiring a barge for access. Similarly there are many other examples that would fall between these extremes.

PRESENT VALUE ANALYSIS

If one once has the data to make proper comparisons as outlined in the previous paragraph, it is also necessary to understand the use of Present Value analysis to account for the time value of money. Neither the concepts nor the calculations are difficult. The calculations can be carried out on many hand calculators.

The results of a Present Value analysis are frequently amazing. For example, assuming a 35 year life for wood poles and 70 years for concrete, using a 10% interest rate and a 5% inflation rate and assuming a tangent structure location in which the labor and equipment cost to change out a pole is only twice the cost of buying the pole, mathematical calculation shows that to produce the same life cycle cost for either a wood pole or a concrete pole line, a 75% premium can be paid for the concrete pole at the time of initial purchase.

If the results of both life cycle costing and present worth concepts were combined, the economic choice for practically all transmission line construction and pole change-outs would be for prestressed concrete poles.

SUMMARY

The concepts in this paper are not likely to find widespread usage among those who place a significant reliance on the "we've always done it that way" theory. A more desirable philosophy would be that expressed by a famous plaque at the University of Wisconsin which reads:

"What ever may be the limitations which trammel inquiry elsewhere, we believe that this great state University of Wisconsin should encourage that continual and fearless sifting and winnowing by which alone the truth can be found."

Hopefully the majority of the utility industry will follow the innovative leaders in that "fearless sifting and winnowing" process. When they do, both structural failures and economic failures of transmission lines will be fewer in number and the use of prestressed concrete poles will increase dramatically.

REFERENCES

(1) Mozer, J.D. <u>Longitudinal Unbalanced Loads in Transmission Line Structures</u>, Documentation of Computer Programs BRODI2 and BROFLX, EPRI Report EL-2943, March 1983.

(2) Peyrot, A.H. and Dagher, H.J. <u>Reliability-Based Design of Transmission Lines</u>, Journal of Structural Engineering, ASCE, Vol. 110, No. 11, November 1984.

(3) Peyrot, A.H. <u>SAPS</u>, A computer program for the static analysis of power and communication systems, 1985.

(4) Peyrot, A.H. <u>Micro-Computer Based Nonlinear Structural Analysis of Transmission Line Systems</u>, IEEE Transactions on Power Apparatus and Systems, 1986.

PRESTRESSED CONCRETE STRUCTURES

Appendix A

FLEXIBILITY ANALYSIS EXAMPLE

STRUCTURE AND CONDUCTOR PARAMETERS

The structure consists of prestressed concrete poles 130 feet long buried 15 feet deep. The structure has a "V" string supporting the center phase and Horizontal Vee suspensions for the two outboard phases. Static wires are attached to the tops of the poles in a conventional manner. The crossbrace is made of steel tube.

Conductor phases are a twin bundle 795 MCM (Drake) and the shield wires are 3/8" EHS. Mounting heights are 100 feet for the conductors (the same height as the attachment of the horizontal strut to the pole) and 115 feet for the statics. Figure 1 shows a schematic of the structure configuration.

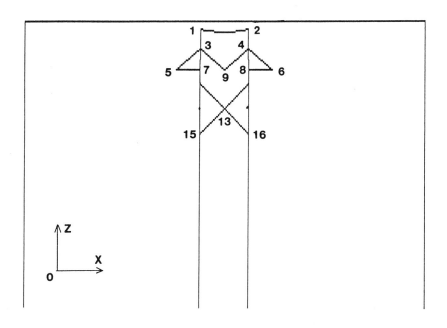

Figure 1

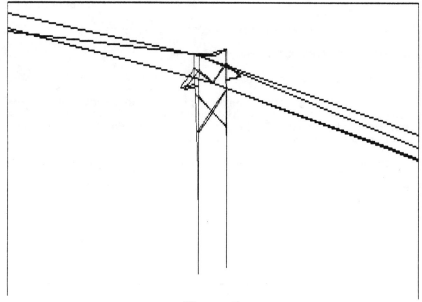

Figure 2

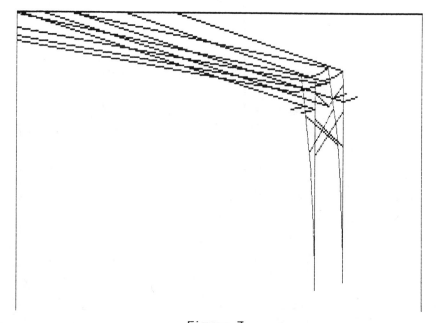

Figure 3

PRESTRESSED CONCRETE STRUCTURES

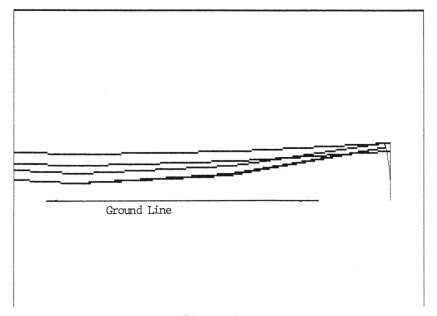

Ground Line

Figure 4

LOADING CONDITIONS

The structure is to be designed for NESC Medium and a broken phase condition with no wind or ice. To illustrate the point of the paper, in the broken phase case, the normal line tension of 15,800 pounds will be applied in the longitudinal direction to an outboard conductor location. No allowance will be made for reduced line tensions due to the flexibility of the system or for the restraint of the remaining intact conductors. Utilizing these assumptions results in a ground line moment requirement of about 1350 ft-kips which becomes the controlling design parameter. The NESC Medium case only requires the poles to be about 600 ft-kips.

By considering the intact restraint and system flexibility, the actual ground line moment requirement is about 550 ft-kips which is considerably less than the results using the incomplete set of assumptions and the broken conductor case is no longer controlling. In fact, the incorrect design assumptions result in a pole that is more than able to withstand all of the wires broken simultaneously. This is not necessarily an undesirable design result, but it is certainly not consistent with the intended results.

To illustrate the point, a run was made assuming that all wires were broken simultaneously. The resultant equilibrium condition dictated by the flexibility of the system results in conductor tensions dropping from 15,800 pounds to about 6,800 pounds and the required ground line moment to withstand the load is only 1,250 ft-kips which is close to, but exceeds the 1,350 ft-kips moment obtained under the original single broken phase condition.

Graphical output from an IBM PC run of the SAPS program shows the proportionally scaled deflections in the system. Figure 2 shows the relatively minor deflection of the structure under the broken phase condition along with the changed conductor sags due to a redistribution of conductor tensions. Figure 3 shows that under a condition of all broken wires, the structure flexes much more, but that the deflections are certainly not unreasonable. In figure 4 it is shown that although there is a considerable increase in sag due to the reduced line tensions, even under this extreme case, the conductors do not touch the ground.

OBSERVATION

The only way to be certain of the end result is to analyze the line as a complete and flexible system. Anything less is pure speculation and subject to a high probability of both structural and economic failures.

Cold-Formed Angles for Transmission Towers

By Edwin H. Gaylord,[1] F. ASCE

ABSTRACT: Cold-formed members enable the engineer to design transmission towers that may be more cost-effective than those built with hot-rolled angles. Members can be designed to closely fit the design requirements. Recommendations for the design of cold-formed shapes for towers are presented. Test data provides verification.

INTRODUCTION

Most transmission towers are fabricated from hot-rolled steel angles. However, hot-rolled angles in the thinner sections are no longer readily available, which tends to increase the cost of towers, particularly those for which relatively thin angles produce significant savings in weight. Cold-formed shapes provide an alternative. In addition, shapes that are not or cannot be hot-rolled can be cold-formed, which can result in more economical towers. One such example is the 60° angle, which can be used as posts in towers of triangular configuration; this eliminates the necessity of using gusset plates to connect the diagonals. Another example is the 90° angle with stiffening lips. These lips increase the local buckling strength of the legs considerably and also make the member less susceptible to damage during shipping and erection. These and other examples of cold-formed shapes that can be used in towers are shown in Fig. 1. Some of these shapes, used as legs, enable

Fig. 1. Typical cold-formed shapes.

longer unsupported panels to be used, which reduces the number of bracing members and bolts.

The Specification for the Design of Cold-Formed Steel Structural Members (10) provides recommendations for design of cold-formed members. However, with a few modifications, ASCE Manual 52, Guide for Design of Steel Transmission Towers (6), can be extended to apply to axial compression of shapes such as those of Fig. 1 and, in addition, to the lipped angle in compression with the framing eccentricities normally found in towers. Design formulas in Ref. 10 include a factor of safety, and where they are quoted here they have been adjusted to conform to the ultimate-strength procedure used in transmission-tower design.

[1]Prof. Emeritus of Civil Engrg., Dept. of Civil Engrg., Univ. of Illinois at Urbana-Champaign, Urbana IL, 61801.

AXIAL COMPRESSION

Formulas for compression members in Manual 52 are those recommended by the Structural Stability Research Council (7). The allowable unit stress (ksi) on the gross section of axially loaded members is

$$F_a = \left[1 - \frac{1}{2}\left(\frac{KL/r}{C_c}\right)^2\right] F_y \qquad \frac{KL}{r} \leq C_c \qquad (1)$$

$$F_a = \frac{291{,}000}{(KL/r)^2} \qquad \frac{KL}{r} \geq C_c \qquad (2)$$

$$C_c = \pi\sqrt{\frac{2E}{F_y}} \qquad (3)$$

in which F_y = minimum guaranteed yield strength of the material; E = modulus of elasticity = 29,500 ksi (203 x 10³ MPa); KL/r = largest effective slenderness ratio of any unbraced segment of the member.

1. Doubly symmetric and point-symmetric open cross sections. Members with this type of cross section fail by overall buckling in one of three modes: bending about either principal axis, or by rotation about the longitudinal axis (torsional buckling). The allowable stress for torsional buckling may be determined by substituting for r in Eqs. 1 and 2 the equivalent radius of gyration r_t given by (1, 4)

$$r_t = \sqrt{\frac{C_w + 0.04J(KL)^2}{I_{ps}}} \qquad (4)$$

where C_w = warping constant, J = St. Venant torsion constant, K = effective-length coefficient, L = unsupported length, and I_{ps} = polar moment of inertia of cross section about shear center.

Torsional buckling is seldom significant since only relatively short members fail in this manner, and even then the critical stress is not significantly smaller than for flexural buckling.

2. Singly symmetric open cross sections. Members with this type of cross section have only two overall buckling modes: (a) bending about the axis normal to the plane of symmetry of the cross section and (b) a combination of torsion with bending about the axis of symmetry, which is called torsional-flexural buckling. The u-axis is taken to be the axis of symmetry in this paper. The allowable stress for torsional-flexural buckling may be determined by substituting for r in Eqs. 1 and 2 the equivalent radius of gyration r_{tf} given by (1, 4)

$$\frac{2}{r_{tf}^2} = \frac{1}{r_t^2} + \frac{1}{r_u^2} + \sqrt{\left(\frac{1}{r_t^2} - \frac{1}{r_u^2}\right)^2 + 4\left(\frac{u_0}{r_t r_u r_{ps}}\right)^2} \qquad (5)$$

where r_t = equivalent radius of gyration for torsional buckling (Eq. 4), r_u = radius of gyration about u-axis, u_o = distance between centroid G and shear center S, r_{ps} = $\sqrt{I_{ps}/A}$ = polar radius of gyration about S, $I_{ps} = I_u + I_z + Au_o^2$, I_u = moment of inertia about u-axis, I_z = moment of inertia about z-axis, and A = area of cross section.

In general, torsional-flexural buckling of plain angles need not be checked. This is because, for the relatively short members for which the lateral-torsional buckling stress is less than the flexural buckling stress for the z-axis, the difference between local buckling of the legs and torsional-flexural buckling of the member is relatively small (1). However, this is not true for lipped angles.

3. Unsymmetrical open cross sections. Members whose cross sections have no axis of symmetry have only one overall buckling mode, which is a combination of torsion with simultaneous bending about both principal axes. The critical value of the axial compression P is given by a cubic equation (12), from which a cubic equation for an equivalent radius of gyration may be derived.

4. Closed cross sections. Members with closed cross sections (box sections and the like) are so torsionally stiff that only buckling by bending about the principal axes need be considered.

EFFECTIVE-LENGTH COEFFICIENT

For lateral-torsional buckling K in Eqs. 1, 2 and 4 is K = 1 if the member is free to warp at each end and to rotate about the u-axis at each end. If warping and u-axis rotation are prevented at both ends, K = 0.5; if they are prevented at only one end, K = 0.7. Mixed end conditions can be treated by replacing r_t and r_u in Eq. 5 with r_t/K_t and r_u/K_u, and K in Eq. 4 with K_t, where K_t amd K_u are the effective-length coefficients for torsional and u-axis buckling. Equation 5 in this form gives the value of K/r_{tf} by which L is multiplied for use in Eqs. 1 and 2. However, K = 1 should be used in Eq. 4 when it is used in conjunction with the effective lengths specified later in this paper.

LOCAL BUCKLING

If they are too thin relative to their widths, legs of angles, and webs and flanges of other types of cross section may buckle locally at stresses less than the yield stress. Local buckling of the legs of a plain angle simply supported and free to warp at both ends is shown in Fig. 2a. Each leg buckles in a single wave with the unsupported edges bowing in the same direction, that is, both clockwise or both counterclockwise looking along the member axis. Local buckling of the leg of a lipped angle is multiwaved in the longitudinal direction (Fig. 2b).

Elements with only one of the edges parallel to the direction of the compression supported, as in the plain angle, are called *unstiffened elements*, while those with both such edges supported, as in the lipped angle, are called *stiffened elements*.

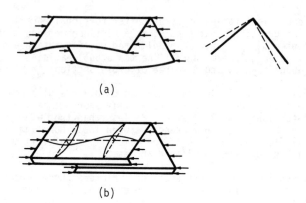

Fig. 2. Local buckling of angles in compression.

Unstiffened elements. If the slenderness of an unstiffened element exceeds the value given by

$$\left(\frac{w}{t}\right)_{lim} = \frac{79}{\sqrt{F_y}} \tag{6}$$

where w = flat width and t = thickness of the element, the element will buckle locally at a stress less than the yield stress and F_y in Eqs. 1 and 3 must be replaced with F_{cr} as given by the following equations

$$F_{cr} = \left[1.659 - 0.659 \frac{w/t}{(w/t)_{lim}}\right] F_y \quad \left(\frac{w}{t}\right)_{lim} \leq \frac{w}{t} \leq \frac{144}{\sqrt{F_y}} \tag{7}$$

$$F_{cr} = \frac{9500}{(w/t)^2} \quad \frac{w}{t} > \frac{144}{\sqrt{F_y}} \tag{8}$$

It is recommended that the width used to determine w/t be based on a maximum inside bend radius of two times the material thickness. A larger radius can be used in fabrication.

Equations 6-8 are compared in Fig. 3 with the corresponding formulas of Refs. 6 and 10. The limiting value of w/t given by Eq. 6 is the same as that of Ref. 6 but larger than that of Ref. 10. Values of F_{cr} by Eqs. 7 and 8 are larger than those of the corresponding equations of Ref. 6. Values by Eq. 7 are somewhat larger than those by the corresponding equation of Ref. 10 for $63.3/\sqrt{F_y} <$ w/t $< 106/\sqrt{F_y}$ but smaller for w/t $> 106/\sqrt{F_y}$. Equation 8 gives values about 30% smaller than the Ref. 10 values for single angles and as much as 50% smaller for other shapes. Thus, values of F_{cr} by the formulas suggested in this paper are conservative compared to the AISI values except for the range of w/t from $63.3/\sqrt{F_y}$ to $106/\sqrt{F_y}$. It is shown later that the formulas are in good agreement with results of tests on hot-rolled and cold-formed plain angles.

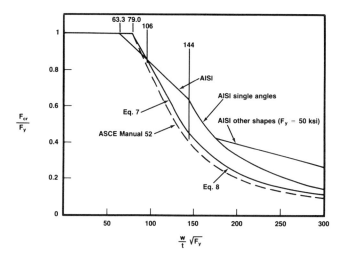

Fig. 3. Local-buckling stress predictions compared.

Equations 1-3 apply to members with stiffened elements only if the values of w/t of the stiffened elements do not exceed the value given by

$$\left(\frac{w}{t}\right)_{lim} = \frac{221}{\sqrt{F_y}} \qquad (9)$$

This limiting value is the same as that of Ref. 10. If $(w/t)_{lim}$ is exceeded, the element will buckle locally at a stress less than the yield stress. Formulas for evaluating the post-buckling strength of such elements are given in Ref. 10, but since elements with $w/t > 221/\sqrt{F_y}$ are likely to have limited application for transmission towers they are not repeated here.

The dimensions of a lip which supports an edge of a stiffened element should be determined as follows (10):

$$d_{min} = 2.8t \sqrt[6]{\left(\frac{w}{t}\right)^2 - \frac{4000}{F_y}} \geq 4.8t \qquad (10)$$

Equations 6-8 have been compared with tests only on plain and lipped angles. Equation 8 may be too conservative for other shapes since it gives values of F_{cr} considerably below those of the AISI Specification. Test results for other shapes are available and should be used for further investigation of these formulas.

MEMBER END CONDITIONS

The effects of member end connections on the buckling strength of plain angles are covered in the ASCE Guide (6) by specifying values of KL/r to be used in Eqs. 1 and 2. These values were based on a review of formulas that had been used by the industry for many years, many of which were supported by full-scale tower tests. Values to account for partial fixity of members with large slenderness ratios ($L/r > 120$) were based on the AISC Specification (9) for bracing members and on tests by the Bureau of Standards (11). The recommendations of the ASCE Guide follow.

For leg sections or post members, bolted in both faces at connections, $K = 1$. For all other compression members carrying calculated stress, the following values of KL/r should be used:

For members with concentric loading at both ends of the unsupported panel, $KL/r = L/r$ for values of L/r up to and including 120.

For members with concentric load at one end and normal framing eccentricities at the other end of the unsupported panel,

$$\frac{KL}{r} = 30 + \frac{0.75L}{r} \qquad 0 < \frac{L}{r} \leq 120 \qquad (11)$$

For members with normal framing eccentricities at both ends of the unsupported panel,

$$\frac{KL}{r} = 60 + \frac{0.50L}{r} \qquad 0 < \frac{L}{r} \leq 120 \qquad (12)$$

For members unrestrained against rotation at both ends of the unsupported panel, $KL/r = L/r$ for $120 < L/r < 200$.

For members partially restrained against rotation at one end of the unsupported panel,

$$\frac{KL}{r} = 28.6 + \frac{0.762L}{r} \qquad 120 < \frac{L}{r} \leq 225 \qquad (13)$$

For members partially restrained against rotation at both ends of the unsupported panel,

$$\frac{KL}{r} = 46.2 + \frac{0.615L}{r} \qquad 120 < \frac{L}{r} \leq 225 \qquad (14)$$

A single-bolt connection is not considered as offering restraint against rotation. A multiple-bolt connection offers partial restraint if the connection is to a member having adequate flexural strength to resist rotation of the joint. Points of intermediate support do not offer restraint to rotation unless they meet these criteria.

Tests on plain angles connected by one leg with from one to three bolts are reported in Ref. 8. Test strengths for members with $L/r = 84$

and 122 and with 2- and 3-bolt connections ranged from 95% to 114% of the ASCE Guide predictions based on the measured dimensions of the specimens. For members with L/r = 176 and with 2- and 3-bolt connections the test strengths were from 105% to 146% of Guide values. Agreement for members connected with only one bolt at each end was not as good; test values ranged from 76% to 105% of Guide Values for L/r = 84 and 122 and from 137% to 163% for L/r = 176.

CONNECTION ECCENTRICITY OF LIPPED ANGLES

Bolts on the centerlines of the legs of a lipped angle are eccentric (Fig. 4b). If a concentrically loaded singly symmetric compression member fails in flexural buckling in the plane of symmetry its strength is decreased by eccentricity in either direction along the u-axis. However, if the concentric-load failure is by torsional-flexural buckling the strength increases with eccentricity in the direction of the shear center, except that the increased strength cannot exceed the load which causes yielding under the combination of axial stress and the stress due to y-axis bending (3). For lipped angles of the proportions likely to be used in transmission towers, r_{tf} is less than r_z, so that torsional-flexural buckling is critical. This and the fact that the eccentricity is in the direction of the shear center S and is small relative to the distance between the center of gravity and the shear center, indicate that the recommended effective-length coefficient, K = 1, for plain angles is satisfactory, and perhaps conservative, for lipped angles.

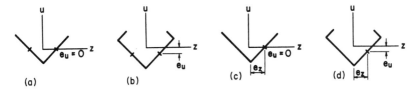

Fig. 4. Connection eccentricities.

Figures 4c and 4d show that, for bolts on the centerline of one leg, the lipped angle has a load eccentricity e_u in addition to the eccentricity e_z common to both the lipped angle and the plain angle. Members with the load eccentric to the plane of symmetry have only one failure mode, namely combined torsion and bending about both principal axes. Calculation of the critical load requires the solution of a cubic equation (10). However, a comparison of a plain angle with a leg width equal to the sum of the stiffener and leg widths of a lipped angle of the same thickness shows the lipped angle to be torsionally stiffer than the plain angle and its maximum extreme-fiber stress, P/A, plus the bending stresses, Mc/I, to be smaller. Therefore, to the extent that the modified effective-length coefficients give good results for plain angles bolted by one leg they should be satisfactory for lipped angles.

TEST RESULTS

In addition to results of tests reported in Refs. 2, 8 and 11, various unpublished tests conducted in 1975-76 on hot-rolled and cold-formed angles at SAE's tower-testing site at Lecco, Italy and tests on full-scale towers were reviewed during the development of the recommendations of this paper (5). Results for hot-rolled angles are compared in Fig. 5 with a nondimensional plot of Eqs. 1-3 and 6-8. The width-thickness ratios were computed using a width measured from the edge of the fillet to the toe of the angle as specified in Ref. 6.

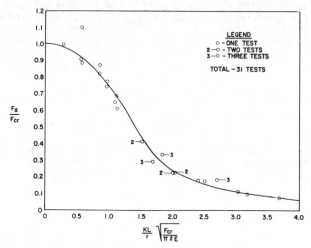

Fig. 5. Tests on hot-rolled angles.

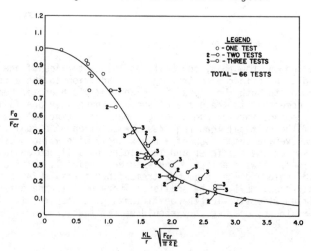

Fig. 6. Tests on cold-formed angles.

Allowable load capacities were computed using the coupon yield values, the measured cross-section dimensions, and where applicable the effective lengths of the ASCE Guide. The coupon yields varied from 37 to 65 ksi (255 to 448 MPa) and the width-thickness ratios from 8 to 17.

Test results for cold-formed plain angles are compared in Fig. 6 with a nondimensional plot of Eqs. 1-3 and 6-8. Allowable load capacities were computed using the coupon yield values, the measured cross-section dimensions, and, where applicable, the effective lengths of the ASCE Guide. The coupon yields varied from 30 to 59 ksi (207 to 407 MPa) and the width-to-thickness ratios from 10.6 to 22.

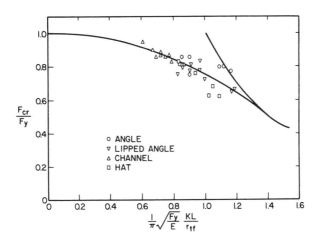

Fig. 7. Tests on miscellaneous cold-formed shapes.

Results of tests (2) on 30 cold-formed plain angles, lipped angles, channels and hat sections in compression are shows in Fig. 7. Thirteen of the specimens were made of 10-gage hot-rolled sheet (thickness = 0.135 in.) and the remainder of 12-gage cold-rolled sheet (thickness = 0.105 in.). Yield strengths, determined by the 0.2% offset method from compression tests of short, laterally supported specimens, ranged from 41.4 to 47 ksi (285 to 325 MPa) for members of hot-rolled sheet and from 30.4 to 36.5 ksi (209 to 251 MPa) for those of cold-rolled sheet. All of these members failed by torsional-flexural buckling. Therefore, the equivalent radius of gyration given by Eq. 5 was used to plot the results. Since the columns were fully restrained at the ends, warping and rotation were prevented and $K = 0.5$. A plot of Eqs. 1 and 2 is shown in the figure.

AXIAL TENSION

Allowable stress for tension is the same in Refs. 6, 9 and 10, except that the AISI and AISC allowables are given as yield stress divided by a factor of safety. Therefore, provisions of the ASCE Guide can be used for cold-formed members.

DETERMINATION OF SECTION PROPERTIES

Evaluation of torsional-flexural buckling involves some properties of the cross section which are not encountered in flexural buckling, among them the torsion constant J, the warping constant C_w, and the position of the shear center. The torsion constant J is given by $J = 1/3 \; \Sigma \; bt^3$, where the summation is for all the elements of the cross section. Procedures for computing the warping constant are described in (1, 12, 13) and other sources. A method for locating the shear center is given in (4, 12) and other sources. Formulas for the shear-center coordinates and for J, C_w and other properties can be found in (1, 3, 13) and other sources.

Formulas for properties of the lipped angle are given in Fig. 8. These are based on square corners. Equations which include the effect of round corners are given in Ref. 3. The formulas can be used for plain angles by setting the lip dimension, d, equal to zero. In general the differences in properties based on square or round corners are not significant (13).

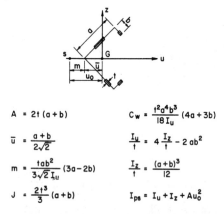

$A = 2t(a+b)$

$\bar{u} = \dfrac{a+b}{2\sqrt{2}}$

$m = \dfrac{tab^2}{3\sqrt{2}\, I_u}(3a-2b)$

$J = \dfrac{2t^3}{3}(a+b)$

$C_w = \dfrac{t^2 a^4 b^3}{18\, I_u}(4a+3b)$

$\dfrac{I_u}{t} = 4\dfrac{I_z}{t} - 2ab^2$

$\dfrac{I_z}{t} = \dfrac{(a+b)^3}{12}$

$I_{ps} = I_u + I_z + A u_o^2$

Fig. 8. Properties of lipped angles.

MATERIAL AND FABRICATION TOLERANCES

There are many suitable steels available for the production of cold-formed members. Ref. 10 provides an extensive listing of steels with yield points from 36 to 65 ksi. Steels with good elongation properties should be used. Material thickness and fabrication tolerances are also given in Ref. 10. Cold-formed members provide the designer with an infinite number of configurations and sizes. However, economic considerations during fabrication dictate judgment in the number of sizes selected for a design.

EXAMPLE

Fig. 7 illustrates the use of the recommendations of this paper to determine the capacity of a lipped cold-formed angle.

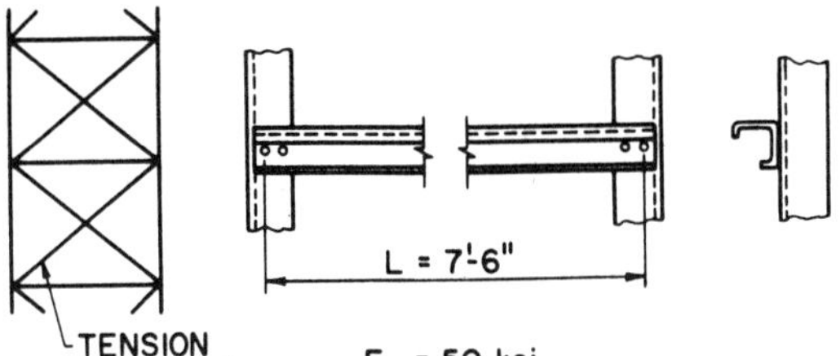

COLD-FORMED ANGLES

$L = 7'\text{-}6''$

TENSION DIAGONALS

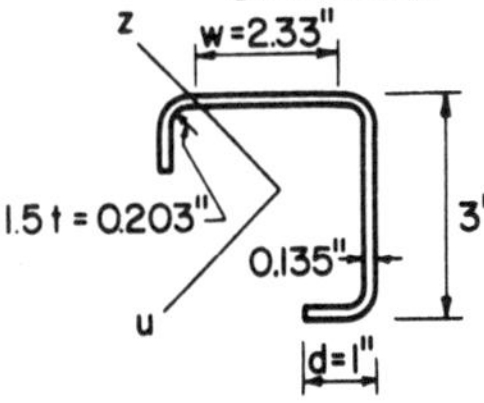

$w = 2.33''$
$1.5t = 0.203''$
$0.135''$
$d = 1''$
$3''$

$F_y = 50$ ksi

$\left(\dfrac{w}{t}\right)_{\text{lim}} = \dfrac{221}{\sqrt{F_y}} = \dfrac{221}{\sqrt{50}} = 31.3$

$\dfrac{w}{t} = \dfrac{2.33}{0.135} = 17.3 < 31.3$

$d_{\min} = 2.8\,t\sqrt[6]{\left(\dfrac{w}{t}\right)^2 - \dfrac{4000}{F_y}}$

$= 2.8 \times 0.135\sqrt[6]{17.3^2 - \dfrac{4000}{50}} = 0.93''$

$A = 1.03$ in.^{2}
$\bar{u} = 1.34$ in.
$m = 0.30$ in.
$u_0 = 1.34 + 0.30 = 1.64$ in.
$J = 0.00623$ in.^{4}

$C_w = 0.441$ in.^{6}
$I_u = 1.79$ in.^{4}, $r_u = 1.32$ in.
$I_z = 0.616$ in.^{4}, $r_z = 0.773$ in.
$I_{ps} = 1.79 + 0.616 + 1.03 \times 1.64^2$
$\quad = 5.18$ in.^{4}, $r_{ps} = 2.24$ in.

$r_t = \sqrt{\dfrac{0.441 + 0.04 \times 0.00623 \times 90^2}{5.18}} = 0.689$ in.

$\dfrac{2}{r_{tf}^2} = \dfrac{1}{0.689^2} + \dfrac{1}{1.32^2} + \sqrt{\left(\dfrac{1}{0.689^2} - \dfrac{1}{1.32^2}\right)^2 + 4\left(\dfrac{1.64}{0.689 \times 1.32 \times 2.24}\right)^2}$

$= 4.903$

$r_{tf} = \sqrt{\dfrac{2}{4.903}} = 0.639$ in. $< r_z$

$\dfrac{L}{r_{tf}} = \dfrac{90}{0.639} = 141$

$\dfrac{KL}{r_{tf}} = 46.2 + 0.615 \times 141 = 133$

$C_c = \pi\sqrt{\dfrac{2 \times 29500}{50}} = 108$

$F = \dfrac{291,000}{133^2} = 16.4$ ksi

$P = 16.4 \times 1.03 = 16.9^k$

Fig. 9. Design of lipped angle.

CONCLUSIONS

Recommendations have been provided for the design of axially compressed cold-formed members of electrical transmission towers, and of plain and lipped angles in compression at the end eccentricities normally found in towers. Cold-formed members provide member size-to-thickness ratios with maximum efficiency. In the development of new tower designs cold-formed members can be produced in many special configurations giving increased flexibility in member selection and significant savings in weight. The number of members and connecting bolts can also be reduced.

APPENDIX - References

1. Bleich, F., Buckling Strength of Metal Structures, McGraw-Hill Book Co., New York, NY, 1952.
2. Chajes, A., Fang, Pen J., and Winter, G., "Torsional-Flexural Buckling, Elastic and Inelastic, of Cold-formed Thin-walled Columns," School of Civil Engineering, Cornell University, Ithaca, NY, August 1966.
3. Cold-formed Steel Design Manual, American Iron and Steel Institute, Washington, D.C., 1980.
4. Gaylord, E. H., and Gaylord, C. N., "Design of Steel Structures," 2nd ed., McGraw-Hill Book Co., New York, NY, 1972.
5. Gaylord, E. H., and Wilhoite, G. M., "Transmission Towers: Design of Cold-formed Angles," Journal of Structural Engineering, August 1985.
6. Guide for Design of Steel Transmission Towers, Manual No. 52, by a Task Committee on Tower Design, ASCE, 1971.
7. Guide to Stability Design Criteria for Metal Structures, B. G. Johnston (ed.), 3rd ed., John Wiley & Sons, Inc., New York, NY, 1978.
8. Madugula, M. K. S., and S. K. Ray, "Ultimate Strength of Eccentrically Loaded Cold-Formed Angles," Canadian Journal of Civil Engineering, June, 1984.
9. Specification for the Design, Fabrication and Erection of Structural Steel for Buildings, American Institute of Steel Construction, Chicago, 1978.
10. Specification for the Design of Cold-formed Steel Structural Members, American Iron and Steel Institute, Washington, D.C., 1980.
11. Technologic Papers of the Bureau of Standards, No. 218, Department of Commerce, August 3, 1982.
12. Timoshenko, S., and Gere, J. M., Theory of Elastic Stability, 2nd ed., McGraw-Hill Book Co., New York, NY, 1961.
13. Yu, Wei-Wen, Cold-Formed Steel Structures, McGraw-Hill Book Co., New York, NY, 1973.

DESIGN AND TEST OF COLD FORMED STEEL MEMBERS

Adolfo Zavelani, m.ASCE *

INTRODUCTION
Latticed systems represent the most spread and low-cost solution for electrical trasmission towers, when ground conditions do not affect the design choice. Steel angles have been used for these structures since ever, but other solutions were also successfully tested, such as tubular and aluminum sections. The actual trend is still towards open sections and bolted connections. Cold formed sections of various shapes have already proven to match the design needs, as well in full scale laboratory tests and in working trasmission lines.
Advantages that can be obtained depend upon:
- easier supply of rough material, which consists of standard size coils;
- possibility of shaping the sections so as to conform the design needs;
- optimal design of the structure in terms of steel weight, number of members and connections.

In many cases cold formed sections provide outstanding performances, when compared to similar hot rolled sections, thanks also to thickness reduction and size increase, which lead to higher inertia properties (fig.1). However, the adoption of thinner walls involves some extra cautions in the design process.

TORSIONAL MODES
The design under compressive forces of rolled and welded sections made up with flat walls, is a typical problem of steel structures, and is generally covered by standard codes of practice, as long as thinness ratios correspond to standard production shapes. Torsional deflections are often considered not to affect the buckling modes, or, alternately, critical loads corresponding to low slenderness ratios are modified so as to account for torsional modes.
The effects of the higher thinness ratios of cold formed shapes must be explicitly taken into account, when drafting specific design recommendations. Uniformly small thickness means very low De Saint Venant torsional stiffness. For doubly symmetric and point-symmetric shapes this leads to feature a torsional buckling mode which can be compared to the usual flexural modes about the principal axes of inertia. In this case the actual capacity of the compression member corresponds to the least of the three critical loads (fig.2b).

* Professor, Structural Engineering Department, Politecnico of Milano, piazza Leonardo da Vinci 32, 20129 Milano, Italy.

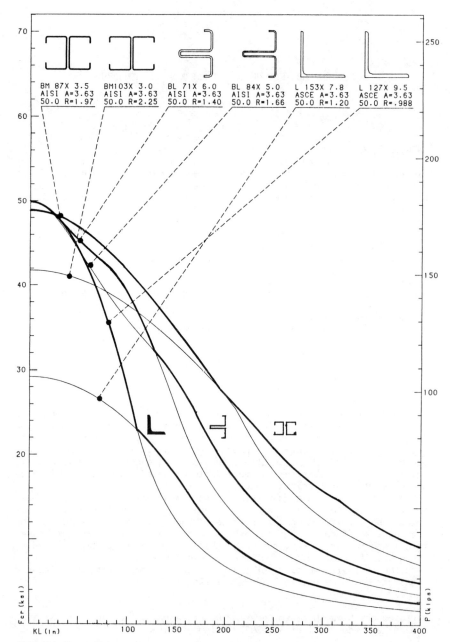

Fig.1 - Capacities of rolled angles (according to ASCE Manual 52) and cold-formed sections (according to AISI Specifications) of equal area.

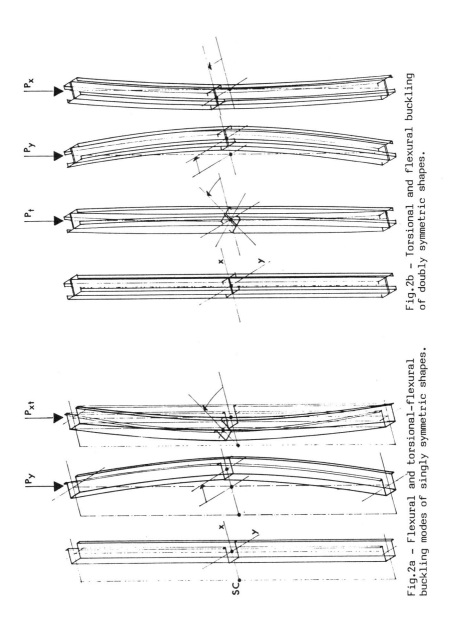

Fig.2b – Torsional and flexural buckling of doubly symmetric shapes.

Fig.2a – Flexural and torsional-flexural buckling modes of singly symmetric shapes.

For sections which are symmetric with respect to the x axis, two modes, torsional and flexural about x, result to be coupled. Two independent critical loads can be separately computed, namely a flexural mode about y and a torsional-flexural mode about x. The member capacity will correspond to the lower of the two (fig.2a).

Non symmetric shapes, which are not particularly appealing for applications to latticed structures, give rise to one only critical load, which results from the combination of the three basic modes.

Formulas for computing the critical loads of cold formed compression members in the elastic range are not reported here for short, recalling that they are easily available in the literature. It is only worth mentioning that three additional section parameters should be computed for the evaluation of the torsional buckling load, namely the position of the Shear Center SC, the Warping Constant Cw and the De Saint Venant torsional costant Jt.

The elastic flexural buckling stress of hinged compression members can be expressed as a function of the only parameter "minimum inertia radius". In order to compare the behaviour of singly symmetric shapes in the presence of torsional modes, also the above parameters must be taken into account. As a consequence, a unified description of the capacity of a family of thin-walled sections can be obtained by representing the critical loads versus the length rather than versus the slenderness ratio, this latter being meaningful only for flexural modes.

In order to size the effects of torsional modes in thin-walled sections, some examples are reported, which provide the critical load curves under the hypothesis of undefinitely elastic behaviour, and comparing equal area sections.

A significant comparison concerns the response of "hat" shapes having different ratios between the flat elements (fig.5). The choice is done so as to obtain circular ellypses of inertia (i.e. equal gyration radii in two directions). The shear center ranges between quite different positions, including both sides of the centroid. The torsional-flexural mode xt is governing all over the considered range of slanderness ratios (up to 240), with respect to the purely flexural modes x,y. For high slenderness ratios of the T-shaped closed hat, the flexural mode about y provides critical loads of the same order of magnitude. This section, having the lowest gyration radius R = 2.2", provides the best perfomances for lenghts over 120" in the elastic hypothesis and over 160" according to AISI formulas. Open hat sections are more sensitive to torsional effects, but still appear to be more effective in the lower slenderness range.

However, the position of the shear center is not the only parameter which influences the torsional-flexural behaviour of the element. Shapes of equal area displaying a similar position of the shear center with respect to the centroid, but different mass distribution, provide different performances, so that each shape appears to be competitive for a particular application range. Similar remarks hold true for plain angles and lipped angles.

LOCAL BUCKLING

Thin walled sections may undergo local buckling, which roughly corresponds to the typical modes of fig.3 for stiffened and unstiffened elements. It is commonly accepted that a wall element is assumed not to displace out-of-plane at one edge (unstiffened) or at two edges

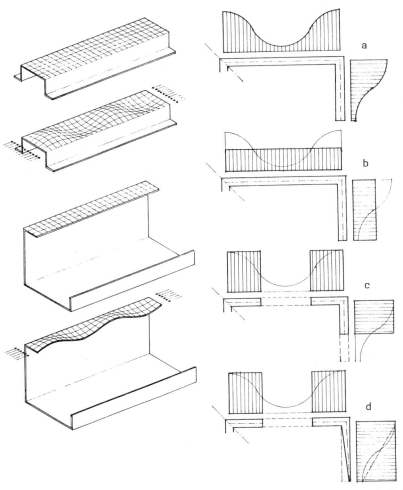

Fig.3 - Local buckling of stiffened and unstiffened flat elements.

Fig.4 - Post-buckling stress distribution on stiffened and unstiffened flat elements (a); average stress on the gross area after local buckling (b); effective areas for the evaluation of the section capacity under ultimate stresses.

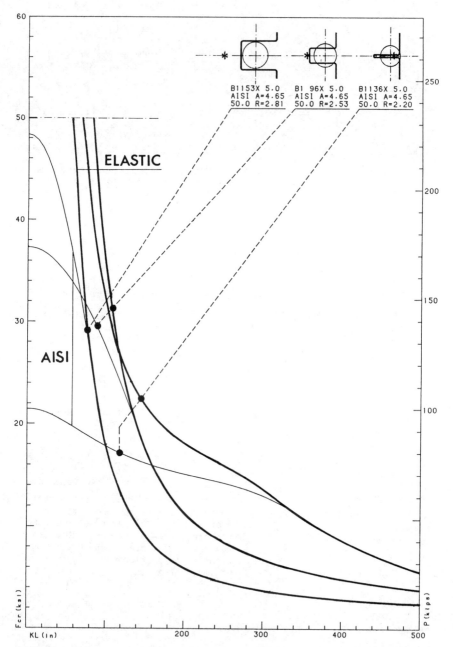

Fig.5 - Comparison of hat sections having equal area and different size ratios: the elastic critical loads are computed neglecting local buckling.

(stiffened). Intermediate situations could be considered, which are not discussed here.
The local buckling stress is generally computed according to the analysis of a rectangular plate, supported at one or two lateral edges, which carries given in-plane loads at the short edges. Minor variations depend on post-buckling effects. The actual post-buckling stress distribution results to be non uniform, with higher values corresponding to the supported edges (fig.4a).
Assuming that the stress peaks cannot overcome the yield point, the section is often remodeled so as to reproduce the above effects in a simplified way, corresponding to a uniform stress distribution.
The AISI approach, as well as most of the current european recommendations, consists in evaluating a mean value of the critical stress depending on local buckling and assuming this last as a limit stress for the gross section. A penalyzing Q factor is thus defined, which cuts the ultimate stress (fig.4b).
An alternative approach consists in defining a new reduced geometry of the section, or "effective section", on which the ultimate stress is considered to be uniformly applied (figs.4c,4d).

DESIGN CRITERIA

Once a factor penalyzing the allowed stress is defined for the section, accounting for local buckling effects, cold formed shapes can be designed according to standard design criteria. Flexural and torsional-flexural buckling modes are taken into account, along with possible end eccentricities, imperfections, second order effects, etc. AISI Specifications reflect to large extent a design approach which is familiar to american steel designers. Similarly, the european codes for cold formed shapes make reference to design curves already applied in steelwork, once the section limits are suitably defined.
On the other hand, the definition of "effective section" is suscetible to introduce an alternative design approach, based on the limit state rather than on the admissible stress concepts. Redefining the geometry of the section allows for better describing the mechanical behaviour of the compression member, even in the post-buckling range. However, it must be noted that the actual deflections of the element depend on the effective section, as well as the effective section depends on the actual deflections. Dealing the problem in terms of effetive sections means introducing an iterative process which can be more accurate and provide a better estimate of the critical load. This promising approach, which appears to be slightly more complex, is being presently investigated by several scientists, and in particular at the Cornell University.

EXPERIMENTAL RESULTS

A confirmation of the expected capacity of elements to be used as compression members of trasmission towers can be obtained alternately on testing machines or in simple structural assemblies.
In the former case, end connections tend to reproduce theoretically defined conditions, say hinged or clamped edges. In the latter, end connections are intended to reproduce actual bolted joints, which are considered to behave as perfect hinges in design assumptions.
When torsional effects are taken into account, theoretical assumptions of hinge restraint impose that end rotations are not prevented and

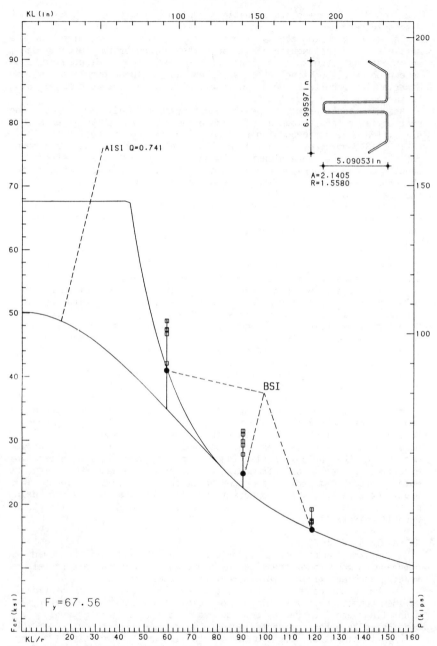

Fig.6 - Experimental results obtained on a testing machine with Cardan joints and ball bearing supports of a 60° T-section subject to local buckling. Expected capacities accordind to AISI and British Standards, and elastic critical curve neglecting local buckling.

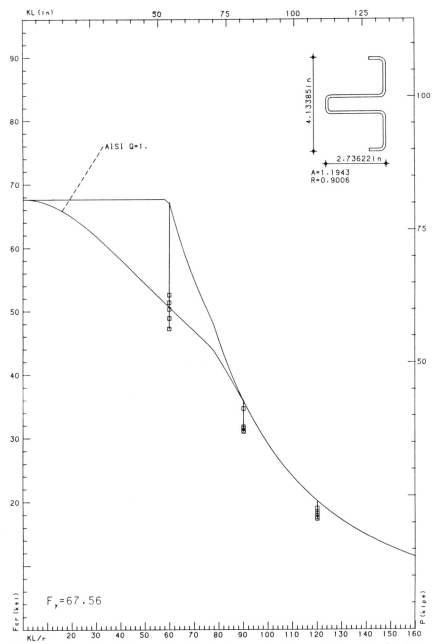

Fig. 7 - Experimental results obtained on a testing machine with Cardan joints and ball bearing supports of a fully effective T-section. Expected capacities according to AISI and elastic critical curve.

that the end sections are allowed to warp. This last requirement cannot be easily fulfilled with actual headings of specimens on testing machines. In spite of this, machines provide more severe and more accurate results in order to check the reliability of proposed design criteria.

In the following, some experimental results are reviewed, corresponding to sections other than angles, which can provide useful information for applications on trasmission towers and can be be designed according to AISI formulas.

Carpena, 1964. A series of tests was carried out in 1964 by Carpena at the SAE Test Station on "open" (60 degrees) lipped and unlipped channels, to be used as post angles for triangular towers. The testing device was a vertical triangular cantilever , and the section to be tested was the vertical chord in compression. Although the behaviour of such shapes is not particularly appealing for practical applications, the results are very interesting, since they closely confirm the validity of the approach.

Fang, 1966. In the framework of the Cornell University Research Project on cold formed sections, several tests were carried out by Pen J.Fang in 1966 for a doctorate thesis. The tests accurately covered the field of axially loaded compression members. The test set-up was a testing machine with flat heads and the specimens were prepared as clamped bars welded and flanged. Warping at the end sections was thus prevented and the expected buckling was theoretically on two fields. The results were published on the Bulletin of Cornell University and were presumably used as a basis for the AISI Specifications of 1968.

Zavelani, 1979. A short series of tests was carried out in 1979 in the Laboratory of the Structural Engineering Department of the Politecnico (University) of Milan, on coupled "open" channels. The test set-up was a MTS displacement driven machine, equipped with greased 4" spheric heads. Flanged specimens provided results which can be conservatively achived as critical loads "on two fields".

Zandonini and Zavelani, 1984. An extensive study was carried out at the Structural Engineering Department of the Politecnico of Milan, which covered the behaviour of different shapes, including rolled angles, subject to concentric and non-concentric loads. The complete set of results was presented and discussed in detail at the ASCE Structure Congress 1984 in San Francisco. The test set-up consisted of a MTS displacement driven machine with cardan joints at both heads. The end sections of the specimens were connected to the joints by means of standard gusset plates having the strictly required number of bolts. From a theoretical point of view the heading system can be interpreted as a flexural and torsional hinge, with some anti-warping effect provided by a double line of bolts connecting the gusset plates. The restraining system seems to reproduce, on the conservative side, the actual behavior of tower assemblies. The results, obtained on bar lengths corresponding to flexural slenderness ratios 60,90,120, were fully satisfactory and shown that AISI Specifications provide a conservative interpretation of the actual behavior, when torsional-flexural buckling governs.

BOLTED CONNECTIONS FOR STEEL TOWERS

Gene M. Wilhoite,[*] Fellow, ASCE

Laced electrical transmission towers require on-site assembly. Consequently, the number of bolts and pieces are important considerations in the total cost of the structures. ASTM A394(1) bolts have historically served as the predominate fastener for these structures. Recent changes in this specification provide an opportunity for using higher allowable design values thus reducing the number of bolts required. Recommended values consistent with the new specification are outlined in this paper.

Introduction

As long as members of transmission structures are connected together at the tower site, bolts will remain an important element of the structure design. Historically, ASTM A394 has been the primary specification for these fasteners.

Through the 1950's, the A394 bolt (60-ksi tensile value) was used with the conventional steels which had yield points from 33 to 42 ksi and tensile values of 60 ksi. However, during the 1960's, conventional steels having yield points of 36 to 50 ksi and tensile values of 70 ksi became available. A394 bolts were still the most predominate fasteners even though some of the tower members had higher tensile values than the bolts. A few utilities began using ASTM A325 bolts on the main legs fabricated from higher strength steels. The designer worried about the erection contractor, Did he see that large insert on the erection drawing showing where high-strength bolts were specified, and if so, did he ensure that the line crew put the bolts with the special head markings in the proper holes?

During this period, the National Electrical Safety Code (4), which serves as an industry standard, specified that the structure will support the assumed loading conditions multiplied by certain overload capacity factors. Under this total load, ". . . the absence of permanent set on the structure indicates that no part has been stressed beyond the yield strength. Allowance should

---

[*]Consulting Engineer, 5533 Pinelawn Avenue, Chattanooga, Tennessee 37411

be made for bolt slip." Using a strict interpretation of this requirement, a bolted connection could not show any bolt or material deformation under full test loads.

In the 1970's, conventional steels became available having yield points of 36 to 65 ksi and tensile values of 80 ksi. A394 bolts with tensile values of 60 ksi still served as the predominate fasteners. Tests on these bolts were showing minimum tensile values of 72 ksi, some 20 percent higher than the minimum. Some utilities began specifying A394 bolts with tensile values higher than 60 ksi. The NESC (5) was changed and specified that the structure will support the designated loading criteria, ". . . without exceeding the ultimate strength."

Today we are still using 36- to 65-ksi yield values for the material. We now have a new bolt specification, ASTM A394-84a (1). The NESC (6) now specifies that the structure will support the designated loading criteria. No reference is made to yield strength or ultimate strength.

Review of this background history is necessary so that we can focus effectively on our past experience and use it as the basis for developing realistic bolt values for new structures. Experience is a valuable friend. Were we naive in the past? No, we established values based on our experience and developed good designs based on the materials available. Consequently, we will concentrate on blending our experience with the extensive testing that has been performed on bolted joints.

Other Considerations

Two other documents need to be included in this discussion. In 1971 the ASCE published a Guide for Design of Steel Transmission Towers (3). A table of bolt spacings and edge distances was included which was suitable for 36- to 50-ksi yield steel and A394 bolts (60-ksi tensile values). No allowable shear or bearing values were included. The document also stated that ". . . the completed tower should support, without permanent set in any member, the loading to which it will be subjected multiplied by the appropriate overload factors." In 1978 ASCE published Design of Steel Transmission Pole Structures (2). Specific recommendations are contained in this document relative to shear and tension values for bolts. Values were established that were ". . . consistent with the practice of prescribing allowable stresses that correspond to the beginning of yield in a member." On formed plate structures, a minimal number of bolts are used in the assembly. Generally, these are high-strength bolts and often are larger than 1-inch diameter. Normally, these bolts are subjected to a combination of stresses. Because of these circumstances, values conforming to yield stresses were specified.

In laced towers, the great majority of bolts function in shear and bearing, and the use of conservative values can greatly increase the tower cost. An ASCE subcommittee is now working on criteria for bolt recommendations, and its recommendations will be an important contribution to our industry.

Special Considerations

Prior to recommending specific values, we must establish the following ground rules.

1. Bolts for laced transmission towers are designed as bearing-type connections. Bolts are normally drawn tight using impact wrenches or hand wrenches. Consequently, the frictional resistance on bare or galvanized surfaces is not adequate to prevent slippage of the joint as the load on the joint approaches 50 percent of the ultimate capacity (generally referred to as the design load times the overload capacity factor).

2. Many utilities use locking devices on tower bolts to minimize vandalism and to prevent loosening of the nuts under vibratory loads.

3. If threaded portions of the bolt are excluded from the shear plane, bolt values should be determined using the nominal area of the bolt and appropriate shear strengths.

4. The minimum tensile stress of the material, F_u, should be equal to or greater than 1.15 times the minimum yield stress, F_y.

5. Dimensions specified are minimum values and must not be underrun by the rolling or cutting operation.

6. Holes are 1/16 in. larger in diameter than the bolts.

7. Bolts range in size from 1/2- to 1-in. diameters.

8. Bolts have square or hexagonal heads with hexagonal nuts.

9. Flame-cut edges can be considered as sheared edges if the torch is machine guided and care is taken that the cut edges are reasonably smooth and suitable for the stresses transmitted to them.

Design Recommendations

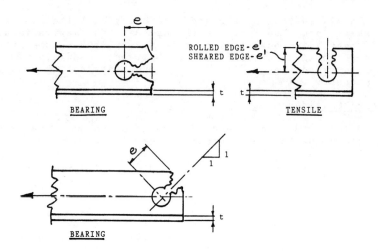

FIG. 1 - MEMBER FAILURE

Fig. 1 shows the type of material failures covered by these recommendations. Tower members are analyzed as pin-connected columns with loads acting perpendicular to the primary axes. Most failures occur due to bearing or tensile stresses.

The recommendations for determining the end distance, e, for a sheared edge are:

For members carrying calculated stress,

$$P_u = \frac{e F_u t}{1.2} \qquad 1.33d \leq e \leq 1.8d \qquad (1)$$

where P_u = ultimate capacity, kips,

e = sheared end distance, in.,

F_u = minimum tensile strength of the material, ksi,

t = thickness of the material, in., and

d = bolt dia., in.

For redundant members, $\qquad e \geq d/0.85 \qquad (2)$

For tensile stresses perpendicular to the stress plane, the following recommendations should be used to determine the minimum edge distance e'.

Rolled edge distance,
(stressed members) $e' = 0.85e$ (3)

(redundant members) $e' = d$ (4)

Sheared edge distance,
(stressed members) $e' = 0.85e + 0.0625$ (5)

(redundant members) $e' = d + 0.0625$ (6)

where e = end distance from Eq. 1, in. and

d = bolt diameter, in.

No fabrication problem exists when using drilled holes in the thicker material with the minimum end and edge distances previously shown.

In normal fabrication of 36-ksi yield material, holes may be punched in members if the thickness of the member does not exceed the hole diameter. (For 50-ksi material, the thickness should be 1/16 in. less than the hole diameter; for 65-ksi material, the thickness should be 1/8 in. less than the hole diameter.)

The minimum end and edge distances in these recommendations may need to be increased when punching the thicker sections. End and edge distances should not be less than $0.9t + 0.5d$ where t = the thickness of the material, in., and d = the bolt diameter, in.

The recommendations for determining the minimum center-to-center hole spacing, s, for stressed members is:

$$s = e + 0.8d \qquad (7)$$

where e = end distance from Eq. 1, in., and

d = bolt diameter, in.

A word of caution: for ease of assembly, a minimum spacing of 3/8 inch plus the long dimension of the nut is required.

The allowable bearing value on the bolts must be checked if the tensile value, F_u, of the member material exceeds the tensile value, F_u, of the bolt. This would occur if a A394, Type 0, bolt with an F_u = 74 ksi is used to connect A572 - Grade 65 material with an F_u = 80 ksi. The full diameter of the bolt should be used for this calculation with an allowable bearing value equal to 1.5 F_u of the bolt. The allowable shear values for the bolts are shown in the ASTM A394 specification as follows:

Type 0 bolts

   55.2-ksi unit shear strength across the area at the root of the thread

   45.88-ksi unit shear strength across the nominal area when threads are excluded from the shear plane

Types 1, 2, and 3

   74.0-ksi unit shear strength across the area at the root of the thread

   74.0-ksi unit shear strength across the nominal area when threads are excluded from the shear plane

Tower bolts are seldom used in tension. Where this is necessary, it is recommended that the allowable stress, F_t, on the net area of the bolt be limited to F_y, the yield stress of the material. If a yield stress is not specified for the material, $F_t = 0.6\ F_u$.

We briefly outlined the design philosophy of transmission towers in past years. Today we expect the structure to support the ultimate loads and remain functional. A member, or a connection, may have a deformation under the ultimate loads. At this time, we have some documentary material to identify what permanent distortion can be allowed in a connection after the tower has supported the ultimate loads. Based upon the values recommended in this document, the following guides are appropriate:

   1. For members: the average distortion of all holes in a connection should not exceed 10 percent of the hole diameter. For example, with two holes, 13/16-inch diameter, the average elongation would be 0.081 inch. If a single hole shows excessive distortion, such as twice the average distortion of the group of holes, the details should be checked to ensure that fit-up was correct.

2. For bolts: the reduction in the net area of the bolt should not exceed 10 percent. This reduction will normally occur at the shear plane.

Summary

The recommendations of this paper reflect the experience of many utilities and extensive testing conducted on bolted joints. The definition of allowable joint distortion under test load conditions requires additional study. This definition must recognize that the structure is functional as long as it can support its intended load conditions and retain the electrical circuits in a safe position. Some distortion and permanent set must be expected in the joints under the ultimate load conditions.

References

1. American Society for Testing and Materials, ASTM Designation: A394-84a.

2. Design of Steel Transmission Pole Structures, American Society of Civil Engineers, 1978.

3. Guide for Design of Steel Transmission Towers, American Society of Civil Engineers, Manual and Reports on Engineering Practice, No. 52, 1971.

4. National Electrical Safety Code, United States Department of Commerce, National Bureau of Standards Handbook 81, issued November 1, 1961.

5. National Electrical Safety Code, 1977 edition, Institute of Electrical and Electronics Engineers, Inc.

6. National Electrical Safety Code, 1984 edition, Institute of Electrical and Electronics Engineers, Inc.

Design of Drilled Piers Subjected
to High Overturning Moments

Anthony M. DiGioia, Jr., F. ASCE[1]

This paper presents the results of a research study directed at improving the state-of-practice, and resulting foundation economy, in the design of laterally loaded drilled pier foundations subjected to high ground-line moments. A theoretical four-spring, subgrade modulus model was developed in this study and was subsequently modified using the results of 14 full-scale load tests. The resultant semi-empirical nonlinear model formed the basis for the development of a design/analysis computer program (PADLL) for drilled pier foundations embedded in multi-layered subsurface profiles. Potential cost savings associated with piers designed using the semi-empirical model are discussed.

Introduction

The drilled pier is a commonly used and cost effective foundation to resist the high overturning moments and shears produced by a variety of loading conditions for single shaft or H-frame transmissions structures. However, since the combination of a high overturning moment and small vertical load is not a common mode of loading, the development of an analytical model to predict the load-deflection relationship and ultimate load capacity of a drilled pier has received limited attention by researchers. Because of the increased use of drilled piers by the electric utility industry in recent years, Electric Power Research Institute (EPRI) funded a research project to develop an improved design methodology for laterally loaded drilled piers.

This paper describes the results of this research effort and is divided into the following major sections:

o Present Practice in the Utility Industry

o The EPRI Research Program

[1] President, GAI Consultants, Inc., 570 Beatty Road, Monroeville, PA 15146

well defined fully plastic region of the load-deflection behavior, identifying a true ultimate capacity was not reached for most of the test piers, where the maximum rotation for the 14 tests was 12 degrees.

Two degrees of rotation of the top of the piers appeared to be a reasonable way of defining the ultimate lateral capacity of these test piers.

Design Model. Referring to Figure 1, the model developed as part of this project is known as a four-spring subgrade modulus model. Translational springs are used to characterize the force-displacement response of the soil; vertical side springs are used to characterize the vertical force-vertical displacement response at the perimeter of pier; a base shear translational spring is used to characterize the horizontal shearing force-base shear response, and a base moment spring is used to characterize the base normal force-rotation response. Details of the model development are summarized in EPRI EL-2197, Volumes I and II.

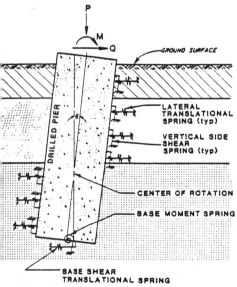

Figure 1. Four-Spring Subgrade Modulus Model

Schematic representations of the force-deflection curves of the four springs are shown in Figure 2. For the piers tested in this research program, the side and base springs were shown to be very important.

Present Practice

Design Philosophy. According to present practice, the transmission line designer identifies a number of extreme load events. The designer must establish the size of foundation which will safely resist the loads associated with these loading events. The designer may also specify how much foundation deflection and rotation are tolerable under these extreme load cases as well as under sustained loads associated with these load cases.

Ultimate Lateral Load Capacity Models. Ultimate lateral load capacity models are used to ensure that the foundation can safely resist extreme load events. Models proposed by Brinch Hansen [2] for multi-layered subsurface conditions and Broms [3] for uniform sand or clay conditions are widely used for the design of transmission line structure foundations. In addition, Matlock [9] has proposed an ultimate lateral pressure formulation for piles in cohesive soil and Parker and Reese [10] have proposed ultimate lateral pressure equations for piles in sand.

Ultimate lateral load capacity for drilled piers is normally computed by assuming that all soil resistance can be characterized by lateral forces on the pier perimeter. However, there is in reality a complex distribution of stresses acting on the perimeter and base of the pier. These stresses result in net vertical forces, in addition to net lateral forces, acting on the perimeter of the pier, as well as a net force and moment acting on the base of the pier. In this regard, Ivey [8] has developed an ultimate capacity model which more closely reflects these forces. Ultimate lateral pressures in Ivey's model are based on Rankine-like passive pressure coefficients.

Linear Load-Deflection Models. A number of linear, lateral subgrade modulus models, e.g., [1, 3, 6, and 12] have been proposed for calculating the lateral deflection of piers and piles. In the subgrade modulus approach, the soil is idealized as a continuous sequence of independent springs, as in the well-known beam on elastic foundation problem addressed by Hetenyi [7]. Using this idealization, the lateral contact pressure, p, at any given depth can be uniquely related to the lateral pier deflection, y, at that depth as follows:

$$p = k_h y \qquad (1)$$

where k_h is the coefficient of horizontal subgrade reaction in units of kips/ft^2/ft of deflection. The following equation was proposed by Terzaghi [12] to relate k_h to the modulus of elasticity of the soil:

$$k_h = \frac{E_p}{1.35B} \qquad (2)$$

where E_p is the modulus of elasticity of the soil and B is the diameter of the pier. This equation, as well as a nominal set of k_h values for granular and cohesive soils proposed by Terzaghi [12], have frequently been used in the utility industry to compute the lateral deflections of drilled piers.

<u>A Nonlinear Load-Deflection Model</u>. Reese and his co-workers at the University of Texas have proposed nonlinear soil p-y curves for slender piers and piles embedded in various soil types [11]. A secant to some point on the p-y curve would correspond to a secant value of the subgrade modulus (k_{hs}).

<u>Generalizations Relative to Current State-of-Practice Techniques</u>. Comparisons between current state-of-practice techniques and the results from full scale tests available prior to the EPRI research project indicated some trends relative to the ability of current state-of-practice techniques to predict the behavior of laterally loaded drilled piers. Some general conclusions in this regard, considering all test data available prior to the EPRI tests, are presented below.

o The load-deflection behavior of laterally loaded drilled piers is highly nonlinear, and thus a linear model cannot predict deflection over a wide range of loads.

o Brinch Hansen's model appears to provide a good prediction of the ultimate capacity of piers embedded in cohesive soil but may underestimate the ultimate capacity of stubby piers embedded in granular soil.

o Broms' model appears to underestimate the ultimate capacity of piers embedded in cohesive soil while over-estimating the ultimate capacity of piers embedded in granular soil.

o Terzaghi's linear load-deflection model considerably overestimates the deflection of laterally loaded drilled piers.

o Reese's nonlinear load-deflection model appears to overestimate the deflection of laterally loaded drilled piers.

The design consequence of the above conclusions are as follows. Piers designed for a permissible lateral deflection using Terzaghi's or Reese's model are larger and thus more expensive than required. Piers designed for an ultimate capacity criterion may be over-designed or under-designed depending on the soil type, the model used

in design, and the size of situation, Electric Power R extensive research program t for designing drilled pier f turning moments.

EPRI Research Program

<u>Field Testing Program</u>. sive data on the field pe subjected to high overturni full-scale prototype tests, c Research Institute and 15 u ducted at various sites acros

Loading of the test pie 80-foot high transmission-li cable and dozer. This arr applying a high ground-lin reasonable vertical and s foundation load was determin and a mechanical dynamomet loading system.

A majority of the test ultimate lateral geotechnica (2700 kN-m). Structural desi test piers was such that geo prior to structural failure. (1.5 m) in diameter and was e ground. Depth to diameter r from 2.5 to 4.2. Details c found in EPRI Report EL-2197,

The following general results of the 14 prototype research project.

1. The test piers b bodies. The cent piers were typica depth below ground to three quarters)

2. The relationship b moment and deflect the pier was highl;

3. Unloading, even moments as compa yielded a consider deflection.

4. At a ground-line the ultimate momen tion was on the or deflection (varied

5.

6.

Propo analytical referred Lateral tr lateral fo side shear shear stre meter of used to c displaceme characteri Details of Report EL-

Figure 1.

Schema relationshi For the 14 shear and

These springs in total contributed from 20 to 40 percent of pier foundation stiffness and between 10 and 35 percent of ultimate lateral capacity.

In order to model pier flexibility, a one-dimensional finite element model was developed consisting of a series of finite beam elements with lateral springs representing the lateral soil resistance and rotational springs representing the vertical side shear resistance. A base shear spring and base moment spring are also used at the bottom node of the model. The stiffness matrix for the beam element was developed using the minimum potential energy theory [4] and an assumed cubic polynomial for the element displacement. The lateral and vertical side shear springs

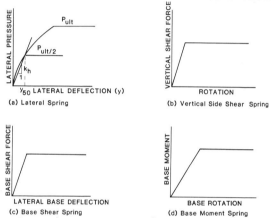

Figure 2. Schematic Representation of Non-linear Springs

were introduced into the element stiffness matrix in a manner consistent with minimum potential energy theory and the cubic polynomial.

Using the semi-empirical, nonlinear four-spring model described above, a design/analysis computer program known as PADLL (Pier Analysis and Design for Lateral Loads) was developed. PADLL can treat both flexible and nearly rigid piers embedded in multi-layered subsurface profiles. PADLL incorporates the four-spring model in combination with the finite beam element model for the pier. Analysis options include ultimate capacity analysis and nonlinear load-deflection analysis. In addition, the computer program can design (select depth and diameter) a pier to satisfy one or more performance criteria.

<u>Comparisons of Measured and Predicted Pier Behavior Using PADLL</u>. Table 1 and Figure 3 summarize the comparison of predicted groundline deflection versus measured deflection.

Table 1

STATISTICAL SUMMARY OF PREDICTED VERSUS
MEASURED DEFLECTION

Load Level in Percent (1)	Error in Percent (2)	
	Mean	Standard Deviation
25	+67	113
50	+25	76
75	+22	83

1. Percent of ultimate capacity as predicted by the four-spring model
2. $\dfrac{\text{Predicted} - \text{Measured}}{\text{Measured}} \times 100$

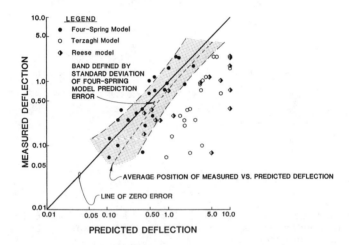

Figure 3. Measured versus Predicted Deflections

Table 2 summarized the comparison of predicted to measured ultimate capacities.

With regard to the comparisons made in Figure 3 and Tables 1 and 2, it may be concluded that the four-spring non-linear model which was calibrated relative to the 14 piers tested as part of the EPRI research project, yielded good predictions of the behavior of these piers.

<u>Potential Cost Surveys</u>. EPRI has published a Technology Applications statement [5], relating the experience of Jersey Central Power and Light Company, which indicates savings of approximately $1,000,000 on future 500 kV lines (approximately 80 miles of line) using the results of the EPRI research program.

Table 2

RESULTS FROM THEORETICAL MODEL PREDICTIONS FOR
ULTIMATE LATERAL CAPACITY

Test Pier	Measured Ultimate Capacity[1] (kip-ft)	Lateral Spring Model		Three-Spring Model		Four-Spring Model	
		M_{ult} (kip-ft)	$\frac{Pred}{Meas}$[2]	M_{ult} (kip-ft)	$\frac{Pred}{Meas}$[2]	M_{ult} (kip-ft)	$\frac{Pred}{Meas}$[2]
1	1980	2033	1.03	2749	1.39	2856	1.44
2	1720	2043	1.19	2219	1.29	2275	1.32
3	2590	1026	0.40	1130	0.44	1195	0.46
4	4400	2559	0.58	3312	0.75	3369	0.77
5	**	3922	++	5725	++	5749	++
6	2950	1818	0.62	2466	0.84	2565	0.87
7	2580	2242	0.87	2473	0.96	2534	0.98
8	3780	2023	0.54	3049	0.81	3112	0.82
9	**	5354	++	5955	++	5960	++
10	4300	3009	0.70	4349	1.01	4384	1.02
11	2550	3360	++	4563	++	4654	++
12	3080	1734	0.56	2162	0.70	2253	0.77
13	2200	2942	1.34	3388	1.54	3484	1.58
14	2230	1745	0.78	1874	0.84	1930	0.87
Average ratio of predicted/measured			0.78		0.96		0.99
Standard deviation of ratio of predicted/measured			0.30		0.33		0.33

[1]M_{ult} defined as the ground-line moment at 2 degrees (35 mRad) of rotation at top of pier.
[2]Predicted ultimate lateral capacity divided by measured ultimate lateral capacity,
**Maximum measured pier rotation less than two degrees (35 mRad).
++Not used to evaluate model.
Note: 1 k-ft = 1.356 kN-m.

Of course, it is dangerous to use the estimated savings associated with a few examples to predict future cost savings for the industry at large. However, it is believed that the following conclusions are valid:

o For utilities which have been using only ultimate capacity criterion, installation savings on the order of 15-20 percent are possible.

o For utilities which have been using both ultimate capacity and deflection performance criteria, cost savings of up to 50 percent are possible.

- The four-spring non-linear model is applicable to nearly rigid piers as well as flexible piers.
- The model is applicable to piers loaded by a high ground-line shear and/or a high ground-line moment.
- The model is applicable to piers embedded in multi-layered soils.
- The model and the computer program PADLL can be used to develop deflection predictions of the lateral load behavior of drilled piers having depth to diameter (D/B) ratios greater than two and less than ten, and can be used for ultimate capacity analysis of piers with D/B ratios between one and ten.

Summary

A rational semi-empirical model for predicting the response of drilled piers to either high ground-line moments or high ground-line shear was presented. This nonlinear semi-empirical model is based on a theoretical four-spring model, developed in an EPRI research project, which was modified to fit the results of 14 full scale load tests.

The result of this study have provided transmission line engineers of the power industry a practical computerized design tool (PADLL). This design tool results in accurate predictions of the response (both ultimate capacity and pier deflection and rotation) of drilled pier foundations.

Significant cost savings can be realized by utilizing the results of this research. The magnitude of such cost savings is dependent upon the current state of practice of the various user utilities.

Appendix A - References

1. Baguelin, F., Frank, R., and Said, Y. H., "Theoretical Study of Lateral Reaction Mechanism of Piles," Geotechnique, Vol. 27, No. 3, November 1977, pp. 405-434.
2. Brinch Hansen, J., "The Ultimate Resistance of Rigid Piles Against Transversal Forces," The Danish Geotechnical Institute Bulletin, No. 12, 1961, pp. 5-9.
3. Broms, B., "Design of Laterally Loaded Piles," Journal of Soil Mechanics and Foundations Division, ASCE, Vol. 91, No. SM3, May 1965, pp. 79-99.

4. Cook, R.D., Concepts and Applications of Finite Element Analysis, Wiley and Sons, New York, 1974, Chapter 3 and pp. 96, 256, and 262.

5. "Design Improvements Reduce Cost of Transmission Line Structural Foundations", Technology Applications, EPRI, 0432C.

6. Gambin, M., "Calculation of Foundations Subjected to Horizontal Forces Using Pressuremeter Data," Sols-Soils, No. 30/31, 1979, pp. 17-59.

7. Hetenyi, M., Beams on Elastic Foundations, University of Michigan Press, Ann Arbor, Michigan, 1946, pp. 52-53.

8. Ivey, D. L., "Theory, Resistance of a Drilled Shaft Footing to Overturning Loads," Texas Transportation Institute, Research Report No. 105-1, February 1968.

9. Matlock, H., "Corrections for Design of Laterally Loaded Piles in Soft Clay," Proceedings, 2nd Annual Offshore Technology Conference, Houston, 1970, American Institute of Mining, Metalurgy and Petroleum Engineering, pp. 577-594.

10. Parker, F., Jr., and Reese., L. C., "Experimental and Analytical Studies of Behavior of Single Piles in Sand Under Lateral and Axial Loading," Research Report 117-2, Center for Highway Research, The University of Texas at Austin, November 1970.

11. Reese, L. C., and Welch, R., "Lateral Loading of Deep Foundations in Stiff Clay," Journal of Geotechnical Engineering Division, ASCE, Vol. 101, No. GT7, July 1975, pp. 633-649.

12. Terzaghi, K., "Evaluation of Coefficients of Subgrade Reaction," Geotechnique, Vol. 15, 1955, pp. 297-326.

IEEE/ASCE TRANSMISSION STRUCTURE
FOUNDATION DESIGN GUIDE

By Paul A. Tedesco *
Member, ASCE

The American Society of Civil Engineers' (ASCE) Subcommittee on Foundation Design for High-Voltage Transmission Structures was formed in October 1976. The purpose of the Subcommittee was to prepare a guide for the design of foundations for electrical transmission line structures. The subcommittee was to review and recommend design methods consistent with the current state-of-the-art for designing various foundation types.

In January 1977, the Institute of Electrical and Electronic Engineers' (IEEE) Working Group on Loading and Strength of Transmission Line Structures appointed a subgroup to prepare a similar guide.

The Chairmen of the ASCE subcommittee and the IEEE subgroup, with the permission of their respective societies, agreed to form a Joint Committee to pursue the objectives and thus avoid the duplication of effort and possible conflicting information which could result from individual efforts. The guide has been submitted to IEEE's Standards Board for approval. Present plans call for the guide to be published by IEEE as a Trial-Use Standard sometime during 1985. A new ASCE Committee has been appointed under the Technical Council on Codes and Standards to review and revise the guide so that the document will be published under the sponsorship of both IEEE and ASCE.

Outline of Guide
The guide consists of the following eight sections:
- Introduction and System Design Considerations
- Loading and Performance Criteria
- Subsurface Investigation and Selection of Geotechnical Design Parameters
- Design of Spread Foundations
- Design of Drilled Shaft Foundations
- Design of Pile Foundations
- Design of Anchors
- Load Tests

This paper will discuss the contents of each of the eight sections of the guide.

* Gibbs & Hill, Inc.
11 Penn Plaza
New York, New York 10001

1. Introduction and System Design Considerations

Electrical transmission line structures are unique compared to other structures in that no human occupancy is involved and the performance requirements are different than for other structure types. In addition, transmission line structures and their foundations may be constructed hundreds or thousands of times in a multitude of subsurface conditions encountered along a line route. Therefore, optimization of design is highly desirable.

Many alternate approaches can be used for the design of foundations for transmission line structures. The guide provides approaches to foundation design that are consistent with the current state-of-the-art, reflect sound engineering practice and successful service experience.

Scope. The material presented in the guide generally pertains to conventional transmission line structure foundations and covers loads encountered as well as foundation performance criteria. Geotechnical considerations, load testing, and construction methods associated with each foundation design are discussed. The guide does not cover the structural design of the foundations or the design of the structure.

System Design Considerations. A transmission line is a system comprised of interconnected elements, each individually designed. Every decision made for the system should consider total installed costs, of which foundations are a major consideration. For example, wire tensions are sometimes increased to minimize the number and/or height of the supporting structures. However, if there are a significant number of angles in the line, total installed costs may be increased due to higher structure and foundation costs at the angles.

Similarly, when developing structure configurations, a wider base may be considered in order to reduce the foundation loads thereby decreasing the foundation cost. This must be evaluated against the added cost of widening the structure.

When designing a transmission line, the engineer has the option to custom design each foundation for a specific site or to develop a minimum number of standard designs that can be used at a majority of the sites. The preferred approach is one that will minimize the total installed cost of the foundations, as well as the line, and may require a combination of both of these methods. The advantages of customizing versus standard designs is further explored and discussed in the guide as they pertain to the various foundation types.

2. Loading and Performance Criteria

Each utility normally has its own unique loading agenda along with corresponding overload capacity factors for the design of transmission line systems. The structural system is analyzed to arrive at foundation loads. Foundations may be designed to have the same strength as, or less or greater strength than, the structures they support depending upon the philosophy of the design team.

For foundation design, the four following types of loads are recognized and discussed in relation to the permeability of the soil in which the foundation is built:
- Steady State - Loads imposed for a long or continuous time period
- Transient - Short duration loads
- Construction - Loads during structure erection or during wire stringing
- Maintenance - Those loads resulting from maintenance activity

Foundation performance is further discussed with relation to the type of structure supported, i.e., lattice tower, single shaft, framed structures, guyed structures, etc. For each structure type, foundation performance criteria must be established by the design team to determine the definition of foundation failure. In essence, failure occurs when pre-established performance criteria are exceeded. Criteria could be magnitude of displacement which would impair the operation of the transmission line or soil stress which would cause foundation failure.

Each structure type used for transmission lines results in different modes of loading applied to the foundations. This and the tolerance of foundation deflection are discussed in the guide for the various structure types:

3. Subsurface Investigation and Selection of Geotechnical Parameters

In order to design safe and cost-effective foundations for transmission structures a thorough knowledge of the subsurface conditions along the right-of-way is required. This section provides a guide for performing adequate subsurface investigations. The objectives of the investigation are to determine the stratigraphy, physical characteristics, and the engineering properties of the soil and/or rock underlying a given site.

The scope of a subsurface investigation is discussed in three distinct phases.

Preliminary Investigation. The purpose is to obtain sufficient data for preliminary engineering; evaluation of the geotechnical conditions for route selection, environmental impact and construction problems, and establishing the basis for development of a detailed exploration program. The preliminary investigation uses considerable existing data coupled with field reconnaissance and a limited boring program. The results of this phase along with economic considerations determine the extent of the remainder of the subsurface investigation.

Design Investigation. This phase provides the information needed for final selection and design of foundations and evaluation of potential construction problems. It utilizes direct methods such as borings, test pits and probes, such as the pressuremeter, as well as indirect methods such as seismic refraction, electrical resistivity and gravimetric surveys.

Construction Verification. Because of the extent of a transmission line and the generally limited number of borings taken, it is advisable that the owner have a representative in the field during construction to determine if actual subsurface conditions are similar to those used in design. If significant variation is found, foundations should be modified accordingly.

This section also presents a condensed discussion of the procedures for taking borings and extracting disturbed and undisturbed samples. In general, disturbed samples and standard penetration test results are sufficient for cohesionless soils, while clays and silts generally require undisturbed samples for measurement of strength and compressibility. For locations where rock is encountered, the pertinent data to be obtained and procedures for rock coring are provided.

The classification of soil and rock samples by visual description and manual tests is an important step in a subsurface investigation program. Generally, based upon the visual classification of soils, a series of index property tests are performed which further aid in classification of the material into categories and assist the engineer in deciding what additional field or laboratory tests will be required.

Index properties for cohesive and cohesionless soils are summarized below:

Cohesionless	Cohesive
Grain Size	Water Content
Specific Gravity	Degree of Saturation
Relative Density	Atterberg Limits
Unit Weight	Specific Gravity
Degree of Saturation	Void Ratio
Standard Penetration Test	Undrained Strength (Pocket Penetrometer or Vane Shear Device)

Index properties are more economically and quickly measured than engineering properties. They are useful because they can be roughly correlated with engineering properties.

Engineering properties are measured directly through laboratory testing. Brief descriptions for the various tests used to determine the shear strength, load-deflection characteristics, and compressibility of soil are discussed in the guide.

In some cases, in-situ tests which measure the engineering properties of subsurface materials may be valuable for designing transmission structure foundations. The more common types which are discussed in the guide are the vane shear, pressuremeter, and plate load test.

4. Design of Spread Foundations

The spread foundation is suitable and commonly used as support for lattice transmission towers. Less common applications are for single shaft and H-framed structures. Several spread foundation types which are frequently used are steel grillages, pressed plates, poured concrete and precast concrete. A description of each of these foundation types is presented in the guide.

This section presents several methods for estimating the uplift and compression capacities and the settlement of spread foundations. The methods are applicable to all types of spread foundations.

Uplift Capacity. The following methods for calculating the uplift capacity are discussed.
- Earth Cone Method
- Shearing or Friction Method
- Meyerhoff and Adams Method
- Balla Method
- Matsuo Method

Each of these methods calculates the uplift capacity on the basis of an assumed failure surface varying in shape from a truncated pyramid or cone to a logarithmic spiral.

Recent studies performed for EPRI by Cornell University provide a review of the methods of calculating uplift capacity and these results derived from an extensive load testing program will be considered in the guide.

Bearing Capacity. The bearing capacity of a transmission structure foundation is defined as the load required to cause a shear (bearing capacity) failure of the supporting soil.

The bearing capacity of the shallow foundation is given in the guide by Hansen's equation.

Settlement. Foundation loads can produce settlement which is the result of immediate or assumed elastic settlement, long-term or consolidation settlement, and secondary settlement or creep. Each type of settlement is discussed with relation to the types of loads caused by transmission structures and methods for calculation are provided.

Contact Stresses. There is, at present, very little information available concerning the response of a spread foundation subject to combined axial forces, large shear forces, and large overturning moments. Methods of analysis by three-dimensional finite element solution and the assumption of elastic springs as supports for the foundation are discussed briefly and appropriate references are presented. A simplified method of computing foundation-soil contact stresses in which the foundation slab is considered to be infinitely rigid and the soil subgrade is linearly elastic, is discussed in detail in the guide. For a a great majority of spread foundations this assumption yields reasonable results.

Construction Considerations. Spread foundations subjected to uplift depend upon the compaction of the backfill. This and other construction considerations are also discussed in the guide.

5. Design of Drilled Shaft Foundations

Drilled shaft foundations have been used successfully to support various types of transmission structures. This type of foundation supports vertical compression loads through a combination of side shear and end bearing, and supports vertical uplift loads by side shear. Lateral loads and overturning moments are supported by lateral resistance of the soil and/or rock in which the shaft is embedded plus the vertical shearing resistance on the perimeter of the shaft, the horizontal shear on the base, and the base moment.

Types of Drilled Shafts. The guide describes the use of three types of drilled shaft foundations: straight and belled drilled concrete shafts, direct embedment of structure shafts, and precast-prestressed hollow concrete shafts. The drilled concrete shaft is the most common type of drilled shaft foundation presently being used to support transmission structures.

Compression Capacity. The ultimate compression load capacity of a drilled shaft foundation is given by the sum of the ultimate end bearing capacity of the base and the ultimate vertical shearing resistance capacity due to skin friction around the perimeter and along the shaft. Equations for determining these parameters are given in the guide.

The ultimate load theory is currently accepted in industry because of its ease of application. However, ultimate load theory does not provide any information concerning the displacement of the shaft as a function of load. Compressive load capacities, as well as load-displacement behavior of shafts, can be obtained by applying modern numerical techniques, such as the finite difference method and the finite element method. However, the sophistication and expense of these methods usually limits their application to special problems and research. Their use requires an electronic computer, and a detailed presentation of these methods is beyond the scope of the guide. However, the guide references technical literature which summarizes these techniques.

Uplift Capacity. There is considerable disagreement concerning ultimate uplift capacity theory due to the problems associated with predicting the geometry of the failure zone. Three of these geometries are commonly referred to as the truncated cone, the curved surface, and cylindrical shear models. A rigorous solution for ultimate uplift capacity is not available at the present time due to the complex behavior of soil under tension loading; however, the guide discusses the present state-of-practice in determining the soil properties and parameters required in the equations for ultimate uplift capacity presented in the guide. The Cornell EPRI Report provides an excellent summary of the methods for determining the uplift capacity of drilled shafts and will be added to the guide.

Lateral Load Capacity. The response of a drilled shaft to lateral loads is the result of complex interactions between the shaft and the soil and/or rock in which it is embedded. A common method of modeling this intereaction is called the subgrade modulus approach. A four-spring subgrade modulus model was developed under an Electric Power Research Institute Research Project. The computer program PADLL which was developed is now available from EPRI for design and eliminates the simplifying assumptions associated with prior models.

The guide presents equations and figures to determine the ultimate load capacity and ground line deflections of both rigid and flexible shafts embedded in either an all-granular or all-cohesive subsurface condition. These charts were developed by Broms and Brinch Hansen.

Direct Embedment. The response of direct embedment foundations including precast hollow shafts to compression, uplift, and lateral loads is similar to that of drilled concrete shafts. As outlined above, most of the analytical techniques used in drilled shaft design are relevant. The principal differences between direct embedment foundations and drilled concrete shaft foundations are: 1) the backfill which intervenes between the pole and the in-situ soil; and 2) the stiffness of the embedded structure shaft relative to that of a drilled concrete shaft. Drilled shafts transfer loads directly to the in-situ soil. However, direct embedment foundations transfer loads to the backfill which transfers the loads to the in-situ soil.

The compression and uplift capacities of direct embedment foundations are also significantly influenced by the type and degree of compaction of the backfill material. For cohesive soils, the undrained shear strength and, consequently, the adhesion between the structure shaft and the backfill will be directly related to the degree of compaction. When granular soils are used as backfill material, the frictional resistance along the structure shaft will depend on the coefficient of lateral earth pressure. If the backfill is compacted to a density comparable to the in-situ soil, the coefficient of lateral earth pressure will approach the at-rest value of the in-situ material. A lesser degree of compaction will result in a value of the lateral earth pressure coefficient which is less than the in-situ value.

6. Design of Pile Foundations

Piles are normally used to transmit loads through soft soils to denser underlying soils or rock. They provide high axial load capacity and relatively low shear or bending moment capacity. Therefore, pile foundations are normally used more often for lattice towers, which have low shear and high axial loads, than for H-framed structures or single shaft structures which have high moments and shear loads.

Since piles are not commonly used to support transmission line structures, they will not be discussed further in this paper, however, they are fully covered in the guide.

7. Design of Anchors

Anchor Types. An anchor is a device which will provide resistance to an upward (tensile) force. An anchor may be a steel plate, wooden log or concrete slab buried in the ground, a deformed bar or a steel cable grouted into a hole drilled into either soil or rock, or one of several manufactured anchors which are either drilled or rotated into the ground. Anchorage may also be provided by vertical or battered drilled shafts or piles.

Anchors may be classified either as dead man or prestressed. Dead man anchors are those which are not loaded until the structure is loaded. Prestressed anchors are loaded to specified levels during installation of the anchor.

Anchor Design. The design of an anchor depends upon a knowledge of the peak and residual shear strength properties of the soil or rock in which it is embedded. In rock, it is also important to know the degree and depth of any weathering which may have occurred, together with the orientation and spacing of joints and foliation. In addition, an understanding of the load characteristics and the structure deflection tolerance combined with the guy cable elongation are important in selecting and designing the type of anchor. Anchor pull-out tests are often conducted to confirm design assumptions where prior experience is lacking.

Grouted Rock Anchors. The ultimate uplift capacity of grouted rock anchors is determined by the following critical interfaces:
- Rock mass
- Grout-rock bond
- Grout-steel bond
- Steel tendon and/or connections

Grouted Soil Anchors. Grouted soil anchors transmit uplift loads to the soil by the following mechanisms:
- Frictional resistance at the grout-soil interface
- End bearing where anchors have a larger diameter than the initial drilled shaft diameter

The actual load transfer mechanisms depend upon the anchor and soil type. The guide summarizes the basic grouted soil anchor types and the soils in which they are used. The anchor types presented are:

Low-Pressure
- Straight Shaft Friction (solid stem auger)
- Straight Shaft Friction (hollow stem auger)
- Under-Reamed Single Bell at Bottom
- Under-Reamed Multi-Bell

High-Pressure - Small Diameter
- Nonregroutable
- Regroutable

Helix Soil Anchors. These anchors develop their ultimate uplift capacity from the bearing capacity of each helix, similar to the bell of a grouted soil anchor. The manufacturers of helix-type anchors recommend a correlation between the torque required to install the anchor, the anchor depth, soil type, number of helixes, and its pull-out capacity.

Helix-type anchors are not normally prestressed; consequently, movement of several inches are common at the maximum design load. Testing each anchor to the design load will reduce in-service movement which will occur during loading.

<u>Spread Anchors</u>. Spread anchors develop their ultimate uplift capacity from the deadweight of the anchor plus the resistance of the soil above the anchor. The vertical component of the anchor uplift load may be analyzed by a number of different methods which are presented in Section 4.

<u>Plate Anchors</u>. Design of plate anchors differ from spread anchors because the in-situ strength of the soil can be used in calculating uplift capacity. Reserach has found that the failure mechanism changes from a soil failure resulting in movement of the soil mass above the anchor for shallow (D/B ≤ 3) and medium depth (3 < D/B < 6) anchors to a localized soil failure at greater depth (D/B > 6); where D is the depth of the anchor and B is the minimum plate dimension.

<u>Group Effect</u>. The capacity of group of anchors depends upon the medium in which the anchor is embedded, the anchor spacing, and depth of embedment. Each design should be checked for group failure assuming the material engaged within the perimeter of the anchor group will fail as a single unit. In cohesive or granular soils, the group would be analyzed as a block of soil whose uplift capacity is equal to the weight of soil within the block plus the shearing resistance along the periphery of the block.

8. <u>Load Tests</u>. Transmission line structure foundations are load tested for the following reasons:
- Verification of the foundation design for a specific transmission line
- Verification of the adequacy of a foundation after construction
- Assistance in research investigations

Many load tests have been performed in such a manner that the results are of little value to the engineering profession. For example, the literature contains many examples of load test results which do not include an accurate and complete description of the soil or rock in which the load tests were performed.

This section is intended to guide engineers to develop testing programs which provide a sufficient quantity and quality of information to make the tests more useful to the individual engineer and to the engineering profession in general.

In general, information provided by load tests reduce the uncertainties inherent in the design of foundations, resulting in a more economical, and safe design. A load test program should be justified by a cost/benefit analysis, that is, the expected cost of the load test program should be weighed against the potential benefits of the information obtained from the load tests.

Instrumentation. The type of instrumentation required will depend on the data which must be obtained to meet the needs of the test program. As a minimum, loads applied to the foundation and movements of the foundation should be measured. The necessity for measuring other parameters such as stresses in the soil and foundation, movements of soil and/or rock in the zone of influence of the foundation, and pore water pressures in the soil near the foundation should be evaluated.

Construction of Test Foundation. Details of construction operations should be well documented by the engineer.

Excavations for construction of the test foundations are helpful in accurately determining the subsurface condition at the test site. Photographs of the construction operations and subsurface conditions should be taken frequently.

Test Performance. The loading and unloading schedule should be established in advance of the test. The number of loading and unloading cycles depends on the requirements of the test program. The loads should be applied in increments and readings of the instruments taken during each increment.

Analysis and Documentation. Analyses of test results can be divided into two parts: 1) those performed while the test is in progress, and 2) those performed after completion of the test.

Analyses performed while the test is in progress give an immediate indication of the behavior of the foundation and allow better control of the test program. Plotting measurements in the field can help to point out anomalous readings.

The results of the tests should be interpreted in a manner which satisfies the requirements of the test program. Some tests will require only a simple determination of whether a foundation moved less than an allowable value under the maximum design load. Others will require sophisticated analyses to arrive at a new method of designing a particular foundation. The analyses should consider the actual subsurface conditions at the test site including additional subsurface information obtained during excavation for the foundation.

The behavior of the foundation predicted by widely-used analytical methods should be compared to the actual behavior of the foundation determined on the basis of test results. This comparison should give an indication of the adequacy of a particular design method for the foundation type and subsurface conditions at the test site.

In foundation engineering, the accumulation of experience from full-scale load tests is an extremely important asset. The test report should be presented such that an engineer unfamiliar with the test can easily follow the procedures and the behavior of the foundation and surrounding ground. The technique used to construct and test the foundation should be fully explained.

TLMRF RESEARCH INITIATIVES

Robert A. LeMaster,* PhD

- Abstract -

This paper describes the research initiatives of the EPRI structural development project RP2016-03. The interrelationship of these initiatives with cosponsored research tests performed at the Transmission Line Mechanical Research Facility is discussed. Examples are presented which show how data obtained during cosponsored tests have been used to obtain an improved understanding of the complex behavior demonstrated by electric transmission structures and how this improved understanding is being used to improve structural analysis software.

*Project Manager, Sverdrup Technology, Inc., P.O. Box 884, 600 William Northern Boulevard, Tullahoma, Tennessee 37388

INTRODUCTION

The Electric Power Research Institute's (EPRI) Transmission Line Mechanical Research Facility (TLMRF) has been performing tests on full-scale electric transmission structures for approximately two years. All tests to date have been cosponsored tests between a utility and/or fabricator and EPRI. A cosponsored test involves two or more organizations using a single structure test to pursue similar or separate objectives. The objective of the utility and/or fabricator is usually to demonstrate the ability of the test structure to withstand a set of applied loads. The pursuit of this objective may or may not require instrumentation. The objective of EPRI is to obtain information which will support the TLMRF structural development program RP2016-03. The pursuit of the RP2016-03 objectives usually requires that a significant amount of instrumentation be placed on the structure. The purpose of this paper is to describe how data obtained by EPRI during cosponsorship tests conducted at the TLMRF are being used to obtain an improved understanding of the behavior of electric transmission structures.

The traditional method for evaluating the strength of transmission structures is based on two steps. The first step is to estimate the member loads or stresses resulting from a given set of loading conditions. The second step is to determine the acceptability of a member to resist these loads by using member strength criterion from an appropriate design guide. This approach inherently assumes that the member loads or stresses are not a function of the member strengths. However, the load in a member is influenced by its strength or the strength of members surrounding it--this coupling is normally manifested when the structure is highly loaded.

In general, current research initiatives at the TLMRF are focusing on three main areas. These are 1) improve the ability to compute member loads, 2) obtain improved member strength information, and 3) evaluate advanced analysis techniques which couple the member load and strength information. Structures tested at the TLMRF can usually be classified as either lattice or tubular steel--wood structures have also been tested. Since lattice and tubular structures are distinctly different, they are discussed separately in the remainder of this paper.

Lattice Structures

To date, 57 lattice structure loadcases have been tested at the TLMRF--this does not include loadcases in which the test objective was to fail the structure. During these tests, unexpected failure has occurred on 23% of the loadcases. However, as shown in Figure 1, the average load level at which these unexpected failures occur is 95.4% of the design level. With the exception of one loadcase, the premature failure loadcases have involved either an unbalanced conductor or shieldwire load. In all cases, the failure occurred in the structure at an unexpected location.

FAILURE NUMBER	APPLIED LOAD LEVEL (%)	FAILURE DESCRIPTION
1	96	SHEARED BOLT, PERMANENT DEFORMATION
2	80	LIGAMENT TEAR OUT
3	100	COMPRESSION MEMBER
4	100	COMPRESSION MEMBER
5	100	COMPRESSION MEMBER
6	90	COMPRESSION MEMBER
7	100	COMPRESSION MEMBER
8	100	COMPRESSION MEMBER
9	97	COMPRESSION MEMBER
10	91	COMPRESSION MEMBER
11	96	LIGAMENT TEAR OUT
12	95	SHEARED BOLT, PERMANENT DEFORMATION

TOTAL NUMBER OF LOADCASES = 57
TOTAL NUMBER OF FAILURE LOADCASES = 12
FRACTION OF LOADCASES IN WHICH FAILURE OCCURS = 21%
AVERAGE LOAD LEVEL AT WHICH FAILURE OCCURS = 95.4%

FIGURE 1. SUMMARY OF LATTICE TOWER FAILURES DURING TEST FROM 8-83 TO 7-85

This information suggests that existing design methodology is serving the industry well. It is possible, using standard practice, to construct a structure which satisfies the design load requirements within an acceptable margin. However, the fact that failures or lack of failures occur at unexpected locations suggests that the actual behavior of the structure under complex load situations may not be that well understood. What happens when a similar structure fabricated under different conditions than the test structure experiences a different yet severe load environment?

Test data indicates that there is considerable dispersion between computed member loads and member loads measured during test. Figure 2 shows a histogram of the percent difference error between test and analysis for all data taken at the TLMRF. This data does not include the effect of residual loads created during erection or resulting from previous testing; residual loads can be quite large and do influence a member's strength. The histogram indicates that approximately 25% of all data has been within ±10% of the analysis results. Figure 3 shows the percent difference error between analysis and test for all loadcases associated with a single test as a function of member load divided by member capacity. Note two things: (1) the more heavily loaded members are more accurately computed, and (2) the majority of the members in the structure are not heavily loaded. In general, the strength of lattice structures tested at the TLMRF have been controlled by a relatively few number of members.

TLMRF RESEARCH INITIATIVE

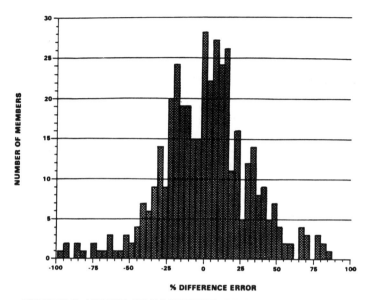

FIGURE 2. ERROR IN COMPUTED MEMBER LOADS FOR LATTICE TOWERS

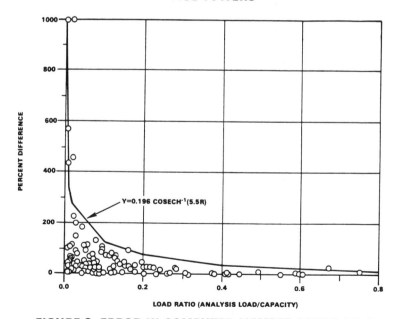

FIGURE 3. ERROR IN COMPUTED MEMBER LOADS AS A FUNCTION OF LOAD SEVERITY

Although the loads in the heavily loaded members such as leg posts are being computed more accurately than lighter members, there is a need to concentrate on the lighter members. Research has shown that error between analysis and test can be significantly improved by improving the agreement between test and analysis in the lighter bracing members. Figures 4 and 5 show the loads in both a tension-only bracing member loaded in compression and an associated leg post. The results of three different analyses are also shown on these figures. The standard tension-only analysis assumed that the member carried no load in compression. The modified tension-only analysis assumed that the member carried its ASCE Guide 52 capacity. The improved tension-only analysis set the member capacity based on the test data. Note that the standard tension-only analysis method over estimated the load in the leg member by 13%, while the analysis based on test data was only 3% in error. By focusing attention on a few bracing members in this test, the agreement between test and analysis was improved in 87% of the results. The 10% improvement in the leg load computation is large enough to permit a decrease in the size of the leg post which would result in a weight reduction for this structure.

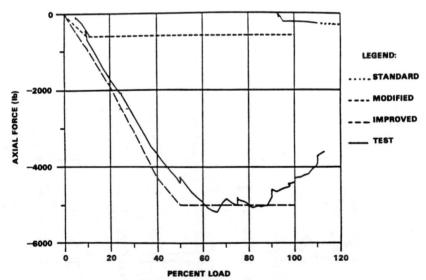

FIGURE 4. COMPRESSIVE FORCE IN TENSION SYSTEM MEMBER

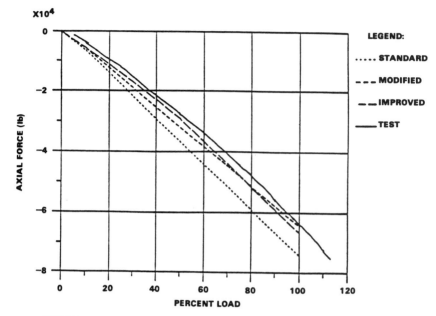

FIGURE 5. COMPRESSIVE FORCE IN LEG MEMBER

In addition to the nonlinear truss element used to perform the analyses in the above example, a nonlinear beam element which incorporates attachment point eccentricities and elastic buckling theory is being used to identify whether improved results can be obtained by modeling more of the detail associated with lattice structures.

Although it is desirable to improve the ability to compute the member loads and capacities, analysis results and design criteria will always have to be interpreted in a statistical sense due to material variability, bolt slippage, erection induced loads, etc. The Transmission Line Mechanical Research Database (TLMRDB) is being created to provide variability information on analysis methods and design criteria. This database provides a standardized storage, retrieval, and sorting system for the large amount of information being obtained at the TLMRF. This database will not only assist the researchers associated with the TLMRF project but will supply the industry with information for improving design guides and implementing LRFD design methodologies. A companion paper presented at this meeting by Dr. Alain Peyrot discusses the TLMRDB in more detail.

Tabular Steel Structures

Tubular steel construction is being used on an increasing basis in overhead electric transmission structures. Research conducted at the TLMRF has demonstrated that when tubular structures remain elastic, nonlinear analysis programs which account for geometry changes as the structure deforms under load can accurately compute the stress distribution throughout the structure. Figure 6 shows the agreement between test and analysis which can be expected when the structure remains elastic.

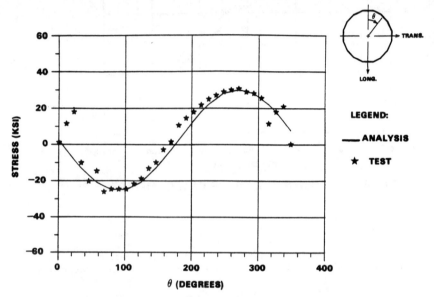

FIGURE 6. AXIAL STRESS DISTRIBUTION AROUND CIRCUMFERENCE OF TUBULAR POLE

The shaft size of tubular steel transmission structures is often controlled by load cases which have a low probability of occurrence, such as the combination of a broken conductor and transverse wind. These low probability of occurrence loads are associated with accident or emergency situations in which the primary objective is to contain failures and minimize damage to the line.

Tubular steel structures used in the electric transmission industry are designed to remain within the elastic limit of the material for all loadcases. Under normal operating loads--those loads having a relatively high probability of occurrence--it is not desirable to exceed the elastic limit of the material. However, for emergency situations, a moderate amount of plastic yielding of the material may be acceptable. It would permit shaft sections to be more compact, which would result in a more flexible line. This increased flexibility would tend to decrease the severity of unbalanced loads, which is beneficial. Although the current design methodology has served the industry well, there may be advantages to having two sets of design constraints, one to deal with normal operating conditions, and one to deal with accident or emergency situations. This is not inconsistent with design methodologies used in other portions of the power industry; the ASME Boiler and Pressure Vessel Codes use a dual standard for steam piping systems.

A nonlinear beam element has been developed to analyze and understand the behavior exhibited by tubular steel structures tested at the TLMRF. The beam element can simulate phenomena associated with large displacements, elastic-plastic material response, and elastic/elastic-plastic buckling. The beam element is applicable to compact closed-polygonal sections, and is operable in the ETAP program. Figure 7 illustrates the geometric properties of the element.

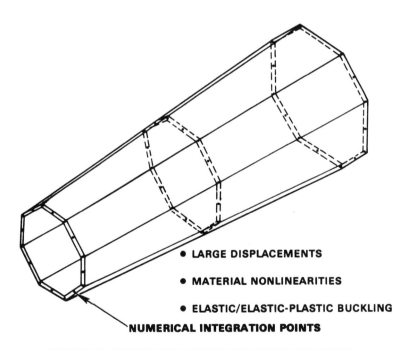

- **LARGE DISPLACEMENTS**
- **MATERIAL NONLINEARITIES**
- **ELASTIC/ELASTIC-PLASTIC BUCKLING**

NUMERICAL INTEGRATION POINTS

FIGURE 7. COMPACT-CLOSED-SECTION-TAPERED BEAM ELEMENT

The element has been used to compute the capacity of a substation take-off structure loaded to 200% of its elastic design load level. In this application, the development of an elastic-plastic hinge in the main cross beam was successfully and accurately computed. Figure 8 shows how closely the analysis followed the test data at one measurement location.

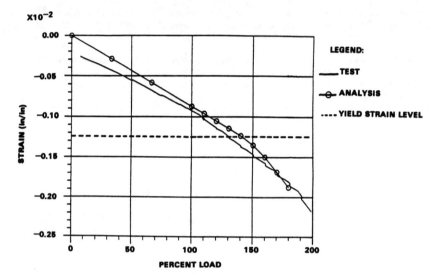

FIGURE 8. STRAIN GROWTH DURING ELASTIC-PLASTIC RESPONSE OF SUBSTATION TAKE-OFF STRUCTURE

The element has also been used to compute the location and load level at which failure occurred in an x-brace member loaded in combined bending and compression during an H-frame structure test. The structure failed at 90% of its design load level under a combined transverse wind and broken conductor loadcase. Figure 9 shows the failed x-brace member.

FIGURE 9. BUCKLED X BRACE MEMBER

Summary

This paper has given an overview of some of the research initiatives being pursued in conjunction with cosponsored research tests performed at the TLMRF. In general, the research initiatives are driven by shortcomings in either existing design criteria or analysis methods which are identified by test results. The research initiatives are broad and seek to improve both analysis software tools and design criteria.

Probably the most innovative or different initiative is associated with looking at the possibility of using the reserve strength or inelastic material behavior of tubular structures. Since this is a new and untried idea, the TLMRF with its test pads, test line corridors, and advanced analysis support capability, provides the laboratory facilities needed to evaluate its potential. In many respects, the TLMRF and its laboratory quality test capability for full-scale structures is possibly the biggest innovation to hit the structures community for a long time.

Acknowledgements

The work reported in this paper was supported by EPRI under project RP2016-03, Mr. Paul Lyons - project manager.

Strain Gaging and Data Acquisition at the TLMRF

Fred Arnold*

Strain gages are the major source of data for the TLMRF research program. Measurement of member loads and bending moments in lattice towers and stress distributions in steel poles are the primary means of evaluating both analysis methods and failure criteria. Gaging procedures, instrumentation, hookup, checkout, calibration, and error analysis will be described as applied at the TLMRF. Because of some unexpected results from the data, calibration and error analysis have become increasingly important to ensure credibility of the data.

Introduction

One of the primary purposes of the Transmission Line Mechanical Research Facility (TLMRF) is to provide experimental data for evaluation of design and analysis methods. Since the designer must predict both the load distribution through the structure and the local load capability throughout the structure, the experimental data required to validate a design method must include both types of data. Thus, an ideal data set for a given structure (lattice or pole) would provide a complete stress distribution throughout the structure for each specified load case and stress values at the point of failure for each load case. The ultimate optimized structure would then be one in which every location is at the point of failure at one or more of the design loads.

Within practical limitations, the TLMRF provides experimental data for selected members or locations in a wide variety of structures during tests. Since strain gages are presently the best choice for quantitative measurements, considerable emphasis has been placed on obtaining accurate strain gage data.

Strain Gage and Data Acquisition Hardware

The strain gages used for most applications at the TLMRF are 350 ohm 1/8 inch (.32 cm) square, temperature

*Operations Manager, Sverdrup Technology, Inc., TLMRF Division, P.O. Box 220, Haslet, Texas 76052.

compensated, polyimide encapsulated gages. Micro Measurement type CEA-06-125UW-350 is the most used single gage. Manufacturers recommendations on surface preparation, bonding, and coating are carefully followed during installation, and all personnel applying gages have completed an application course by Micro-Measurement.

Data Acquisition System

The TLMRF data acquisition system is built around a PDP11/23 computer which is also used as a load control system. In its present configuration, 128 channels are available for all measurements, leaving approximately 80 to 120 channels available for strain gage data depending on the number of load cells required for load application. In this mode, an average of 10 samples is recorded for each channel every 2 seconds.

A block diagram of a typical data channel is shown in Figure 1. Because of the need to record data from any

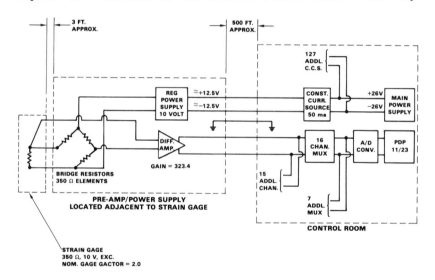

FIGURE 1. SINGLE CHANNEL BLOCK DIAGRAM OF TLMRF DATA ACQUISITION SYSTEM

location on any tower, a design was selected which is insensitive to effects of lead length on excitation voltage and is capable of driving the signal over long lead lengths without significant loss. A combination excitation supply, pre-amp, and line driver was designed which could be located in a small weatherproof enclosure usually less than 3 ft. (.9m) from strain gage locations.

Signal output at the multiplexer input is not significantly affected by line lengths up to 1000 ft. (300m) or more. The use of shielded signal lines operating in a differential mode results in a typical noise of less than 3 millivolts within a set of 10 samples of a 10 volt data channel.

The bridge shown in Figure 1 is for a single gage application and uses 350 ohm precision completion resistors (currently T9 .1%, 25 PPM/degrees C). Thus the strain gage at ±.3% resistance tolerance is the dominant element in determining the zero strain output voltage. Because of the increasing emphasis on total member loads rather than applied test loads, these completion resistors must be replaced to obtain a temperature drift of less than 2 PPM/degree C. This will be discussed further under error analysis.

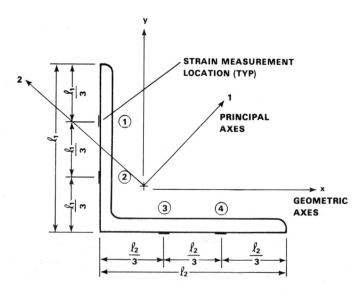

FIGURE 2. AXIAL FORCE AND BENDING MOMENT STRAIN GAGE CONFIGURATION

Lattice Tower Measurements - Combined Force and Moment

Since lattice towers are normally produced from angles, end attachments induce bending moments in the members which greatly influence axial load capacity. At present these moments are not calculated in the design analysis; they are merely allowed for in the code by limiting member capacity as a function of type of end connection and by specifying acceptable types of end connections. If improved analysis methods are to explicitly consider bending moments, then measurements of actual bending moments are required to verify these methods. A four gage arrangement requiring four data channels is used at the TLMRF for combined force and moment measurements in angle members. Figure 2 shows the dimensions of gage locations. Each set of gages is applied as close as possible to the mid-span of an unsupported length. Note that the x and y axes are based on the standard AISC configurations, and that Gage 1 is located on the angle leg parallel to the y-axis.

Strains at the four measurement locations may be written with respect to the principal axis stress resultants as

$$\varepsilon_1 = \frac{1}{E} \left\{ \frac{P}{A} + \frac{M_{11}\ell_{11}}{I_{11}} + \frac{M_{22}\ell_{21}}{I_{22}} \right\} \quad (1a)$$

$$\varepsilon_2 = \frac{1}{E} \left\{ \frac{P}{A} + \frac{M_{11}\ell_{12}}{I_{11}} + \frac{M_{22}\ell_{22}}{I_{22}} \right\} \quad (1b)$$

$$\varepsilon_3 = \frac{1}{E} \left\{ \frac{P}{A} + \frac{M_{11}\ell_{13}}{I_{11}} + \frac{M_{22}\ell_{23}}{I_{22}} \right\} \quad (1c)$$

$$\varepsilon_4 = \frac{1}{E} \left\{ \frac{P}{A} + \frac{M_{11}\ell_{14}}{I_{11}} + \frac{M_{22}\ell_{24}}{I_{22}} \right\} \quad (1d)$$

where ε_i is the axial strain at location i, with i = 1, 2, 3, 4; M_{11} and M_{22} are the bending moments with respect to the principal axes, ℓ_{ij} is the normal distance between the ith principal axis and the jth gage; E is the modulus of elasticity; I_{11} and I_{22} are the second area moments of the cross-section with respect to the principal axes, and P is the internal force acting on the cross-section.

Note in Equation (1) there are four known parameters with three unknown parameters, and the stress resultants cannot be computed explicitly. To compute the best approximation to the three stress resultants based on four strain measurements, a least-squares operation may be performed.

The minimum number of strain measurements required to compute the three stress resultants is three. Thus, the use of four gages at each location provides some redundancy. If one of the four gages has known or suspected errors, that gage is removed from the calculation, and no least squares fit is performed.

In order to maximize confidence in the test data, tensile tests are performed on each gaged member, both before and after a tower test. Prior to assembly and after disassembly each gaged member is installed in a special test setup and loaded in tension. Data is taken as tensile load is cycled from zero to the maximum value 4 times, while load is measured with one of the TLMRF test load cells. Calibration test load is determined by the maximum safe load without special attachment fittings, and ranges from 10% to 25% of yield. This data is then reduced for axial force only, compared with the calculated values, and a force correction factor derived. A practical method of moment application has not yet been demonstrated; therefore, no corrections are applied to moment data. This limited calibration meets the following objectives:

(1) Allows direct correction for angle member dimension tolerances, modulus of elasticity, gage factors, and bridge excitation voltage. The correction is applied to the force data after data reduction.

(2) Provides evidence that all gages and all components of each gage circuit are functioning properly.

(3) Exercises the gage in a manner that shows up major gage slippage problems.

(4) Defines changes in "zeroes" or responsivity during a tower test by comparing pre-test and post-test calibrations.

Data Quality - Combined Force and Moment Measurements

Potential contributors to the total error may be divided into two categories: (1) those errors which affect the overall gage calibration constant (slope of the strain/output curve) and (2) those errors which cause a "zero" shift.

The following factors affect the slope of the strain/output curve either by changing the slope of the best fit linear curve (gage constant) or by causing actual data to deviate from a linear fit:

(1) Linearizing gage and circuit assumptions (1)<1uin/in
(2) Bridge resistance initial unbalance (1)   <.02%
(3) Bridge excitation voltage   ±.25%
(4) Gage Factor   ±.50%
(5) Young's Modulus   ±3-4%

(6) Angle member dimensions ±2.5%

The first four items are negligible in the present application, while items (5) and (6) are accounted for in the axial force measurements by the tensile tests previously described. Tensile test data to date substantiates the preceding values. The slope of the measured force vs. load data agree with calculated values to within 5% with few exceptions. Further, comparison of pre and post test data generally show less than 1% slope change for undamaged members.

The stability of the zero strain output is a different matter. Initially, this was not considered important, since forces in members due to applied test loads were the primary objective, and it was believed that an unloaded tower provided an adequate starting point, either by reading strain at zero load or by extrapolating the member load/tower load curve back to zero applied load. However, it has been found that non-linear and/or history dependent effects may be dominant in lightly loaded members. Examples of those effects are fabrication tolerances, joint slippage under load within bolt clearance holes, and assembly and erection procedures. The load which a member must carry is the total load at any given time, not the change in load due to loads applied to the tower. A stable zero strain reference is as important as a stable known sensitivity. Factors which influence the stability of the zero strain reference are:

(1) Temperature variation of strain gage
(2) Temperature variation of bridge completion resistors
(3) Strain cycling

While errors in sensitivity apply at a constant percentage for all members and loads, zero shift errors result in error percentages which are largest for lightly loaded members.

The effect of temperature variation in the strain gage is well defined in the manufacturers specifications and is correctable if accuracy requirements warrant. In an effort to eliminate errors due to these temperature effects, self-temperature compensating gages are used on all tower tests. When they are matched to the correct base material, these gages are designed to completely compensate for temperature variations over a small range of temperature. There is, however, a very real possibility that the gages used are subjected to temperatures outside the effective range of the self-compensation feature. To demonstrate the possible magnitude of the error caused by temperatures outside this range, a sample gage is selected which was designed to read accurately at approximately 80 degrees F. Figure 3 is a series of curves showing the % error vs. temperature for this gage at different levels of axial stress (f_a). Notice that for members subjected to high stress levels ($f_a = F_y$) the effect of temperature variation in the range shown is minor, but for lightly loaded members ($f_a = 0.125 F_y$) the error can be quite large.

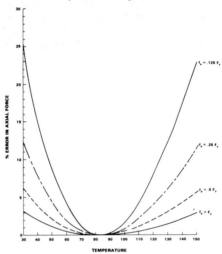

FIGURE 3. EFFECT OF TEMPERATURE VARIATION OUTSIDE COMPENSATED TEMPERATURE ON AXIAL FORCE

Temperature variation of the bridge completion resistors has been a major contributor to observed zero shifts in past TLMRF data. For example, a temperature change of 30 degrees F (17 degrees C) could occur between calibration of a gage set and test data. If we ignore compensation between resistors and assume one resistor has the maximum 25 PPM/degrees C, $\Delta R/R$ is 4.25×10^{-4}. This represents a "zero shift" of .344 volts on the TLMRF data system output or an apparent microstrain of 206 uin/in, equivalent to a stress of 6180 psi. No shifts of this magnitude have been observed, but the uncertainty evidenced by comparing pre-test and post-test zero readings has thus far precluded use of this gage configuration in determining assembly and erection loads. Completion resistors with a temperature coefficient of better than 2 PPM/degrees C will be installed in the near future, and temperature sensitivity of each bridge resistor assembly will be checked. A zero shift error contribution of ±5 uin/in from the completion resistors is believed to be feasible.

Another potential error is the influence of strain cycling. When a new strain gage is loaded, the gage may indicate irregularities due to cold-working of the strain-sensing element. Cold-working of the gage produces resistivity changes in the alloy which are realized in the form of a zero shift in the strain data. In addition to this zero shift, a hysteresis effect may be created where the gage output deviates from a linear relationship with applied strain. The combined effects of the zero shift and the hysteresis effect can lead to strain errors of up to 7% for a typical gage.(2) These effects can be minimized by cycling the gage four or five times before using it to record strains. For the cycling to be effective the gage must be cycled to at least the load to be applied during the test and it must be loaded in the same direction (tension or compression.) The size of the zero shift errors caused by this factor in these tests is not at this time distinguishable from temperature effects; however, no hysteresis has been observed.

Lattice Tower Measurements - Axial Force Only

For members which are designed to carry tension only or where test objectives do not require bending moments, a different gage arrangement is used. This arrangement was suggested by Bonneville Power Administration and is a preferred arrangement if bending moments are not required. Figure 4 shows the gage locations. The four gages are wired in a bridge, and completion resistors are eliminated. For axial force measurements, this arrangement has the following advantages:

(1) Only one data channel required per location
(2) Completion resistors are eliminated
(3) Temperature effects are cancelled
(4) Data reduction is simplified and can be accomplished on-line with manual zero correction

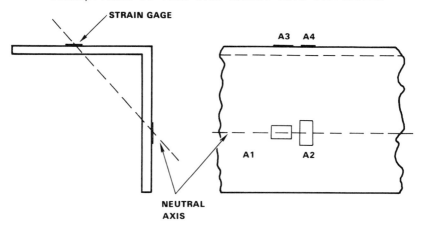

FIGURE 4. AXIAL FORCE ONLY STRAIN GAGE CONFIGURATION

Tensile tests of axial-force-only members are performed in the same manner as for four channel locations. Calibration factors are derived directly from plots of the gage output vs. applied load. Since the maximum calibration load is only 10% to 25% of yield, the upper portion of the calibration plot is used to determine calibration constant (slope) in pounds per volt and zero offset (intercept) in pounds. Comparisons of pre-test and post-test calibrations generally support an accuracy claim of better than 5%. Reliable data on absolute member loads has thus far been obtained only with this arrangement.

Pole Tower Measurements

Strain gage data on steel poles is taken in the same manner as single channel data for lattice members, except that in-place calibration is not an option. As with the four channel angle moment arrangement, the data is of most use in showing changes in stress and stress distributions as a result of an applied tower load. For poles, the importance of history dependent residual stresses is not as great as for lattice structures.

Variations in Young's Modulus must now be allowed, and it is desired to use the results of coupon tensile tests. In the data reduction process, the value of 29,000 ksi (200,000 MPa) is used for Young's Modulus (consistent with the AISC Manual of Steel Construction) and the nominal cross-sectional area of the member is used. According to the ASME Boiler and Pressure Vessel Code the Modulus of Elasticity of low carbon steels (such as A36 or A572) is approximately 28,000 ksi (193,100MPa) at 70 degrees F (21 degrees C). This difference of 1,000 ksi (6.9MPa) in Young's Modulus relates to a 3.5% difference (or error) in the stress calculation.

The location of gages is generally determined by predicted critical areas. The most common use is to ring the lower part of a pole with gages, both on corners and flats (e.g., 24 gages on a 12 sided pole). Other locations used are on H-Frame cross braces, arm attachment points, or any region of concern. It should be emphasized that sufficient gages must be installed to determine stress distribution, since both magnitude and distribution must be compared with analysis.

Planned Improvements

In addition to the bridge circuit improvements discussed earlier, an additional data acquisition system is scheduled for completion in 1985. This system will provide the following:

(1) Increase the total data channels from 128 to 256
(2) Provide on-line data reduction and plotting capability for the original channels plus the additional 128 channels

The system will be a second 11/23 computer with commonality of major components with the existing test control computer. Real time plots, real time reduction of strain gage data including moments, real time deflection monitoring, on-site reduction of calibration data, and long term data monitoring will be added capabilities. The last two items will improve data quality by detecting calibration and stability problems in time for correction prior to test.

An additional improvement now being considered is a method of calibrating a four channel set of angle measurements for bending moments and calculating corrected gage system constants (μin/in-volt) for individual channels.

Properly installed and used, strain gages provide a source of accurate and reliable stress data at a reasonable cost. Our goal at the TLMRF is to be able to provide documented accuracy of better than 5% for strain, forces, and bending moments from assembly through test and disassembly.

REFERENCES

(1) "Linearity and Sensitivity Error in the Use of Single Strain Gages with Voltage-Fed and Current-Fed Circuits", Manfred Kreuzer, <u>Strain Gage and Transducer Techniques</u>, Number 1, Society for Experimental Mechanics, 14 Fairfield Drive, Brookfield Center, CT, 06805.

(2) <u>Experimental Stress Analysis</u>, Dally, J.W., and Riley, W.F., McGraw-Hill, Inc., 1965, 520 pp.

Strength Data Base for LRFD of Transmission Lines

by Alain H. Peyrot[1] , M. ASCE

ABSTRACT

Reliability-Based Load and Resistance Factor Design (LRFD) has been proposed as a desirable alternative or a complement to the National Electrical Safety Code (NESC) for the structural design of high voltage electric transmission lines.

While the LRFD method could already be used to advantage with the existing limited knowledge of the variability of the load and strength variables, the full benefits of the method will only be realized when additional statistical information becomes available. Recognizing the need for better information on loads and resistances, the Electric Power Research Institute (EPRI) has supported and will continue to support research efforts that help better describe these variables. These efforts have already resulted in a comprehensive data base for wood poles.

At the EPRI Transmission Line Mechanical Research Facility (TLMRF), much needed resistance data for components of lattice or other types of structures are already being accumulated. Data from each instrumented or observed component of all the structures tested at the TLMRF become a permanent part of a computerized Data Base (TLMRDB). It is possible to query the TLMRDB to obtain statistical information on structural strength. It is expected that the TLMRDB will become the premier source of strength data for the US power industry.

I INTRODUCTION

Reliability-based Load and Resistance Factor Design (LRFD) has been proposed (5,6,9) as a substitute or complement to existing design methods (7). Advantages of LRFD are simplicity of use, consistency and designer control of the reliabilities of various components in the line.

While the format of LRFD is ideally suited to the design of new lines, other probabilistic methods are available to evaluate the reliability of existing lines that may be candidates for upgrading or reconductoring (5,8).

LRFD combines both load and strength information into a convenient

----

1. Professor, Department of Civil and Environmental Engineering, University of Wisconsin, Madison, 53706

design equation. Considerable information has already been published to help designers select loads, load factors and load combinations, i.e. develop the load side of the LRFD equation (5,6). Strength data are needed to guide the selection of design strengths and strength factors, i.e. develop the strength side of the equation. Some strength information is already available (5), but much more needs to be collected and published to achieve the full benefits of reliability-based analysis or design. This paper describes the type of strength information that is needed.

II FAILURE MODES

There is an almost infinite number of ways in which structural failures can occur in a transmission line. Any failure occurence is a failure mode of the line. Failures can be extensive and involve several structures or they can be limited to one or a few components of a single structure. But in all cases, the failure is initiated by the failure of a single component. The single component failure may be contained (the most desirable outcome), or it may trigger other failures, such as a cascade of structures.

Economy in the construction and operation of a line is achieved by keeping the probability of occurence of its failure modes at appropriate levels. The probability of failure of a single component can easily be calculated. However, the calculation of the probability of occurence of a failure mode involving several components or structures is an extremely complex, if not impossible, task. In such a case, the concept of line SECURITY can be used. Security is a measure of a line's ability to contain initial failures of components. Security can be enhanced by building stronger structures at fixed intervals or specifying sufficient ductility or longitudinal loads to prevent cascading. Once a line security is guarantied, the designer only needs to be concerned with the reliability of individual components. Therefore, component reliabilities (and individual component failure modes) are the main concern of LRFD. Calculations of system reliability can only be rough approximations.

Loads on a transmission line system produce load effects in its components. Detailed data and theories on loads and load effects are included in Refs. 5,6,8 and 9. A component fails if it is stressed beyond its capacity, i.e. if a load effect in the component, Q, exceeds the corresponding strength, R (also called ultimate capacity or resistance). Any separate load combination on the line produces a separate load effect in each component and may stress that component in a different manner. Therefore, depending on how the component is loaded, more that one random variable R may be needed to represent its strengths. For example, an angle member that is subjected to a compression load effect may fail in many complex combinations of yielding, local and member buckling.

Over the lifetime of a line, both Q and R are random variables. The probability of failure of a component, P_f, is the probability that any Q in the component exceeds the corresponding R. The reliability of the component is the inverse of its probability of failure.

It is commonly assumed that a component strength R has a Probability Density Function, PDF, that is Normal (Gaussian) as shown in Fig.1. The PDF is entirely known if two of its parameters are known, say its mean m_R and coefficient of variation, V_R. The component strength that is calculated using a procedure specified in a design guide (1,2,3,4,10) is a single nominal value, R_n. R_n is a conservative estimate by the profession of the true strength R. The true random strength R of a particular component may be below or above its calculated strength. If the probability that R is below R_n is equal to "e", in percent, then R_n is said to be the "e-percent" exclusion limit of strength, denoted as R_e. The exclusion limits corresponding to the R_n values given in current design guides and the coefficients of variations for R are unknown or appear to vary widely. Yet, "e" or both "e" and V_R are needed for a full implementation of LRFD or probabilistic evaluations.

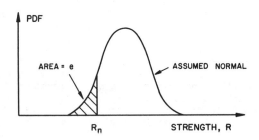

e = EXCLUSION LIMIT (IN PERCENT)

R_n = NOMINAL STRENGTH-FROM DESIGN GUIDE

V_R = COEFFICIENT OF VARIATION

Figure 1. Probability Densisty Function of component strength

Table 1. Strength factor ϕ to adjust component reliability by factor C   e = exclusion limit of R_n

C =	50	10	2	1	1/3
e = 1 pct	.65	.89	1.04	1.16	1.39
5	.56	.77	.90	1.00	1.20
10	.52	.72	.84	.93	1.12
15	.50	.69	.80	.89	1.07

III LRFD FORMAT

The LRFD format (6,9) allows the designer to specify the reliability of any component in a line. That reliability is specified by the selection of a global importance factor, G, and a component factor, C. The product GxC represents the reliability of the component relative to other components in the line or in other lines. The procedure is summarized by the flow chart below. The load side of the LRFD equations (top block of the flow chart including Eqs. 1 and 2) is not emphasized in this paper. The strength side (bottom block with Eq. 3) only includes the nominal strength R_n and an associated strength factor ϕ.

FLOW CHART DESCRIBING LRFD PROCEDURE

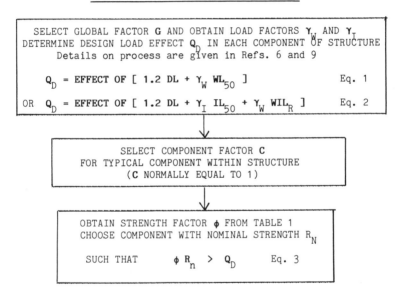

SELECT GLOBAL FACTOR **G** AND OBTAIN LOAD FACTORS γ_W AND γ_I
DETERMINE DESIGN LOAD EFFECT Q_D IN EACH COMPONENT OF STRUCTURE
Details on process are given in Refs. 6 and 9

Q_D = EFFECT OF [1.2 DL + γ_W WL_{50}] Eq. 1

OR Q_D = EFFECT OF [1.2 DL + γ_I IL_{50} + γ_W WIL_R] Eq. 2

SELECT COMPONENT FACTOR **C** FOR TYPICAL COMPONENT WITHIN STRUCTURE (**C** NORMALLY EQUAL TO 1)

OBTAIN STRENGTH FACTOR ϕ FROM TABLE 1
CHOOSE COMPONENT WITH NOMINAL STRENGTH R_N

SUCH THAT $\phi R_n > Q_D$ Eq. 3

Table 2. Strength information needed to implement LRFD

	a)	an estimate of the exclusion limit "e" of the strength R_n (minimum information)
or	b)	estimates of "e" and V_R
or	c)	estimates of m_R and V_R
or	d)	the PDF of R (most complete information)

It should be noted that the value of the strength factor ϕ (from Table 1) depends on two variables: the component factor C and the exclusion limit "e" of the nominal strength R_n. A more accurate value of the strength factor ϕ is available if, in addition to C and "e", the coefficient of variation V_R is known (Ref. 9). Therefore, sizing of a component with Eq. 3 requires that some statistical information on its various strengths be available. The information needed for full implementation of LRFD is shown in Table 2.
All parties concerned with the design of power lines should strive to develop and disseminate the type of information described in Table 2. It is recommended that all future editions of material design guides (1,2,3,4,10) publish strength equations that are at the 5 percent exclusion limit. This would insure that all materials are considered on an equal basis. A side benefit would be to reduce Table 1 to one row.

IV CONTRIBUTING UNCERTAINTIES IN COMPONENT STRENGTH

The true random strength R of a component can be expressed by the equation:

$$R = R_n * r \qquad \text{Eq. 4}$$

where

R_n = nominal strength based on an assumed loading mechanism, the nominal size of the component and its nominal material yield or ultimate stress. R_n is normally obtained from a design guide(1,2,3,4,10)

r = a normalized strength, i.e. the ratio R/R_n

The statistical properties of r can be obtained directly from those of R through the use of Eq. 4.

It is useful to identify the following seven contributions to the uncertainty in r:

E = gross Errors in design, i.e. an error in calculation or conception. Full scale structure testing is in part conducted to eliminate the possibility of serious design errors

M = Material strength uncertainty (yield or ultimate stress)

S = actual Size uncertainty. Fabrication and construction tolerances cannot guaranty an exact size

A = uncertainty due to effect of Aging, i.e. deterioration of strength with time in service

D = uncertainty because of limited Data. Formal statistical theory can be used to quantify the variable D

P = Professional factor to account for the inability of simplified design procedures to predict the true strength even if all over uncertainties in E, M, S, A and D are eliminated

O = Other contributions that may be identified for specific components

The sources of uncertainties described above can be thought of as random variables affecting the normalized strength "r"

$$r = \mathcal{F} [E, M, S, A, D, P, O] \qquad \text{Eq. 5}$$

where $\mathcal{F}$ represents a function (usually unknown).

It is convenient (and often theoretically defendable) to assume that Eq. 5 takes the following simplified form:

$$r = E * M * S * A * D * P * O \qquad \text{Eq. 6}$$

where E, M, S, A, D, P and O are treated as uncorrelated random variables. Statistical properties for E, M, S, A, D, P and O can be obtained experimentally. For example, statistical properties of M can be developed by measuring the ratio of the true yield stress of a steel angle member to its nominal yield stress F_y (F_y = 36 ksi for A36 steel). Properties of S would be obtained by measuring ratios of actual sizes to nominal sizes (nominal sizes are those listed in design handbooks or supplier data). The ratios of a component strength measured after so many years in service to the strength of a new identical component would provide data on A. There are cases, however, where it is not practical to separate all the variables. In such cases, several variables can be grouped and treated as a unit. For example, if material strength and fabrication tolerances are not separated, their combined effect can be represented by a single variable MS. Testing nominally identical short stubs of steel angle members (say 4*4*1/2 angles of A36 steel) would produce data on MS for steel angle members.

The purpose of the professional factor P and the method for gathering statistical data on that factor are best described by the following example. Consider a steel angle (say a main bracing member in the body of a tower) connected at both ends by two bolts in one leg. That angle is subjected to a complex combination of axial load and eccentric end moments. In traditional tower design, the moments due to end eccentricities are not calculated. The tower is modeled as an ideal 3-dimensional pin connected truss. Therefore the load effect that is calculated in each member is a pure axial load, when in fact, the true load effect is a combination of axial load and moment. The details at the end of the member and the geometry of adjacent members all contribute to the magnitude of the moments in the member. The member can fail by combinations of excessive yielding, overall nonelastic buckling controlled by the member slenderness ratios, end restraint and eccentricities, and local buckling of the projecting legs.

ASCE Manual 52 (4) has provisions for determining R_n for steel angle members. The Manual provides equations to guard against the occurence of the failures described in the previous paragraph. These equations have traditionally been applied together with the assumption of no moment in the analysis (truss analysis). These equations assume that the end conditions for the member can be grouped into three simple eccentricity conditions (curves 1 to 3) and three simple restraint conditions (curves 4 to 5).

Data on P could be obtained from measured values of the following ratio:

$$\frac{\text{Measured strength of angle}}{\text{Strength calculated with ASCE Manual 52 using the exact dimensions and material properties measured during the test}}$$

In some cases, data on the combined effect of M, S and P, i.e. MSP could be obtained by measuring the following ratio:

$$\frac{\text{Measured strength of angle}}{\text{Strength calculated with ASCE Manual 52 using nominal dimensions and yield stress}}$$

If Equation 6 is used, then

$$m_r = m_E * m_M * m_S * m_A * m_D * m_P * m_O \qquad \text{Eq. 7}$$

and

$$V_r = \sqrt{V_E^2 + V_M^2 + V_S^2 + V_A^2 + V_D^2 + V_P^2 + V_O^2} \qquad \text{Eq. 8}$$

It can be observed from Eqs. 7 and 8 that approximate statistical properties for r can be obtained from statistical properties obtained separately for E, M, S, A, D, P and O.

If r cannot be expressed by the uncoupled form of Eq. 6, the amount of tests needed to insure that all ranges of values for the variables have been covered can become prohibitive. On the other hand, with the assumption of Eq. 6, each variable can be studied separately with a small number of tests. Therefore, there is an overwhelming advantage in using the model of Eq. 6 when developing strength statistical data.

V EPRI STRENGTH DATA BASE

The TLMRF was constructed to conduct tests to: a) prove compliance of structures with specification requirements, b) validate and improve mathematical modeling techniques for predicting structural failures, and c) develop data bases for various designs and materials.

The first objective meets the immediate need of utilities and fabricators. The second and third objectives are long term. Part of the second objective is to collect data on Analysis uncertainty, i.e. on the ratio of true member load effect to calculated value using specific analysis assumptions. Analysis uncertainty is not discussed in this paper but it is a major item of study at the TLMRF.

Both component and full scale tests are conducted at the TLMRF. From a sophisticated array of data acquisition devices, large amounts of data are collected during each test. To insure that all data will be readily accessible at future times, the Transmission Line Mechanical Research Data Base (TLMRDB) has been created (13). The TLMRDB is a state of the art computerized information management system that is tailored to the needs of the transmission line researcher or designer. Information on each component instrumented or observed during a test is entered in the data base in the form of attributes. The list of attributes depends on the family of components that is considered. For components of steel lattice towers, a list of about 30 attributes allows a complete description of the member, the manner in which it is loaded, its failure mode, etc. The information is sufficient to allow retrieval of statistical data on the variables E, M, S, A, P and O, separately or in groups, as described in Section IV of this paper.

It is reasonable to expect that, within a few years, the TLMRDB will contain enough information to develop the information in Table 2 for typical components. Retrieval and display of the information will be automated. For example, one will be able to request a full display of the PDF of the compression strength of a particular tower member or simply the exclusion limit of the strength of that member calculated by formulae in Manual 52.

VI CONCLUSIONS

The next generation of design procedures for transmission lines will most likely be reliability-based. The theory is already well developed. Because the statistical data that are needed to fully implement reliability-based procedures are often missing, there is an urgent need to collect improved data on loads and strengths. The EPRI testing facility has the capability of generating as well as organizing vast amounts of strength data.

ACKNOWLEDGEMENTS

The LRFD method for transmission lines is based on research sponsored by EPRI and on efforts of the ASCE Committee on Electric Transmission Structures. The TLMRDB was developed and will be maintained for EPRI by Sverdrup Technology, Inc.

APPENDIX A - REFERENCES

1. Committee on Lightweight Alloys, "Guide for the Design of Aluminum Transmission Towers", Journal of the Structural Division, ASCE, Vol. 98, No. ST12, Dec. 1972, pp. 2785-2801.

2. Committee on Steel Transmission Poles, "Design of Steel Transmission Pole Structures", Journal of the Structural Division, ASCE, Vol. 100, No. ST12, Dec. 1974, pp. 2449-2518.

3. Committee on Prestressed Concrete Poles, "Guide for Design of Prestressed Concrete Poles", Journal of the Prestressed Concrete Institute, Vol. 28, No. 3, May/June 1983.

4. Guide for Design of Steel Transmission Towers, ASCE Manual 52, New York, NY, 1971.

5. Guide for Reliability-Based Design of Transmission Line Structures, Research Institute of Colorado, EPRI Research Project RP-1352, 1985.

6. Guidelines for Transmission Line Structural Loading, by ASCE Committee on Electrical Transmission Structures, ASCE 1984.

7. National Electrical Safety Code, ANSI C2, published by IEEE, New York, NY, 1977 and 1981 Editions.

8. Peyrot, A. H. and Dagher H. J., "Theoretical and User's Manual for DESCAL - Reliability Analysis and Design of Transmission Line Components," EPRI Research Project RP-1352-2, August 1984.

9. Peyrot, A. H. and Dagher H. J., "Reliability-Based Design of Transmission Lines," Journal of Structural Engineering, ASCE, Vol. 110, No. 11, November 1984.

10. Specifications and Dimensions for Wood Poles, ANSI 05.1, American National Standards Institute, New York, NY, 1979.

13. Transmission Line Mechanical Research Data Base (TLMRDB), report prepared by Sverdrup Technology, EPRI Research Project RP-2016, 1985.

NESC - A Flexible Document

Richard A. Kravitz*
IEEE Senior Member

The National Electrical Safety Code (NESC) is a living document that has been constantly undergoing modification and growing pains as technology advances.

Some will argue that it has not evolved far enough, and other will argue that it has gone too far.

By its very nature, the NESC document allows innovation in the design of transmission line structures and their foundations. The purpose of the NESC is to define minimum loading requirements and overload capacity factors necessary to protect the public. It is not to provide design guidelines and design procedures, or to recommend structure types and configurations.

Starting in the late 1950's, the trend has been toward construction of high voltage and extra high voltage transmission lines. The early editions of the code, such as the 5th and 6th editions (issued in 1961), included the following excerpted statement in Section 26:

"NOTE: It is recognized that -- and other developments may become available and that ---. It is further recognized that while these materials are in the process of development, they are subject to such test evaluation and trial installations as may be approved ---."

This statement appeared under the section defining ultimate fiber stress of wood poles. However, many engineers used this statement liberally to develop new concepts in structures and composite structures which proved successful over the years.

To reinforce the opinion that the NESC would allow innovation, the 1977 Edition of NESC removed this statement from the wood pole section and inserted it in the first paragraphs of Section 26, thereby recognizing that innovation can apply to all materials and structures.

1. Line Design

In the late 1950's, 345 kV lines were constructed based on experimental work at some test sites. In the late 1960's, 765 kV lines were built based on experimental work at test sites. The clearance requirements used in the design of these projects, including 500 kV, i.e., conductor air gap to structure, to guys, to ground, were based on these test results and not necessarily on extending the criteria given in Section 23 of the code.

*Consultant, Ajikawa Iron Works, 224 Locust Road, Willmette, IL 60091

2. Structure Design

In the 1960's, steel and aluminum guyed towers were starting to be used throughout the world. Canada was using 345 kV guyed vee towers. Guyed "vee" and "wye" towers were being considered for use in the United States. At that time, however the NESC suggested that guys on "metal" structures was undesirable. Utilities considered that this was not really disallowing the use of guys and so guyed metal structures were designed with the guys considered as an integral part of the structure. After the successful installation and operation of many guyed structure installations, the 1977 edition acknowledged the use of guys as an integral part of the structure.

Even though the 6th edition of NESC and the 1973 edition specified the strength requirements for guys, it was interpreted that this data applied to strength requirements of guys used with wood pole structures. Guys considered as an integral part of the metal structure were sized such that the ultimate load in the guys, under NESC loadings and NESC overload capacity factors, would not exceed 90 percent of the rated breaking strength of the guy. Guys used with wood poles would be designed to 37.5 percent of their rated breaking strength (transverse strength) to resist the NESC loadings without overload factors.

Today, guys are used in some structures as replacements for crossarms (chainette or cross rope tower).

In my opinion, there is absolutely no restriction to the use of structures built of composites of metal, plastic, fiberglass or other materials. Certainly, for years, wood pole structures using horizontal post units were used which did not fit any specific category in the older NESC versions.

3. Lattice Tower Design

The majority of steel lattice towers designed in the 1950's utilized the tension-only web system. In this system, diagonal web bracing members were designed to carry axial tension stresses only. This was done by allowing these members to have a slenderness ratio of up to 500, although this would appear to be contrary to the requirements of the NESC. Even the 1973 edition restricted the slenderness ratio of members having stress to 200.

4. Special Conductor Configurations

Recently, a special conductor configuration, consisting of two subconductors wrapped around each other, was developed. The developers have requested that the code specifically denote the method to determine the effective diameter to be used in calculating wind on conductor and ice coating. This is being considered for the next edition of the code.

However, the engineers using this special conductor could also apply the same effective diameters by specifically applying Rule 260B of the latest NESC which recognizes newly developed materials and allows them to be used in trial installations.

5. Clearances

   The 1977 edition of the NESC provided alternate calculations to determine clearance requirements where system switching surges are known and controlled. Many other significant changes were also included in the 1977 edition regarding clearance calculation, such as establishing clearances over several classifications of navigable waterways; including clearances for dc; providing for allowances to correct for altitude changes, and so on.

   In summary, we can see that changes have constantly been made to the NESC. The 1977 edition provided the most extensive revisions and alterations, and the code is constantly undergoing review and revision, and will continue to do so as our engineering profession continues to explore new horizons and innovate new materials and design concepts.

6. Line Design

   See Section 24, Grades of Construction.

   Grade "B" construction is required when a line crosses a railroad. In most all other cases, the latest edition of the code will allow Grade "C" construction or less under all other conditions providing the line can be de-energized initially and following breaker operations, if contact takes place.

   A wood pole line (and if appropriate, a steel structure line), could be built to Grade "C" construction. At critical crossing locations, Grade "B" is required, and Rule 252C1 would apply - that the structures located on each side of the crossing must be designed to carry longitudinal load.

   To provide this longitudinal load, paragraph 261A4 applies. However, to avoid providing Grade "B" construction longitudinal loading at the structures at the crossing, many line designers interpreted Sections 252C and 262A4 such that they could avoid designing for longitudinal load if the change in grade of construction from Grade "C" to Grade "B" were made at a point where Grade "C" construction was required and not Grade "B" and the structures involved were designed for Grade "B" construction transverse and vertical loading instead of Grade "C" loadings.

   Therefore, it was proposed to provide Grade "B" transverse strength structures at the two crossing structures and at both of the structures adjacent to the crossing structures.

   Some may say that this is not innovation, but rather using a loop-hole in the code. In any event, alternate methods to provide required safety to the public can be met by judicious interpretation of the code.

NESC LOADING REQUIREMENTS
Jerome G. Hanson,* M.ASCE

The NESC loading requirements, Section 25 of the code, allows innovations in the Design of Electrical Transmission Structures.

The first Rule of Section 25 states, "In the absence of a detailed loading analysis no reduction in the loading specified therein shall be made without the approval of the administrative authority."

The key to innovation is the word "absence". The code allows decreased loading requirements if the designer does make a detailed loading analysis.

The different loading district boundaries generally follow state or county lines. Everyone knows that the required values for combined ice and wind will not stop or start exactly at these lines. Few utilities have made detailed icing studies. Many utilities know from experience that the required NESC values will not provide reliable lines. Please note, I used the term "reliable". The code provisions are intended to provide safe lines, they will not necessarily result in reliable lines due to local climatic conditions. On the other hand, there are many areas of the country where results from icing studies could be used to reduce code loading requirements and still provide safe and reliable lines.

There are also possibilities for innovations in the extreme wind loading provisions of the code.

The 1977 edition of the code recognized that Medium and Heavy loading on large diameter conductors was non-conservative in high wind areas, and the extreme wind loading was added to correct the deficiency.

The soon to be published 1987 edition will contain a revised wind map; see Figure 250-2. The new wind map is derived from ANSI A58.1-1982 for exposure category C.

The 1987 wind map differs considerably from previous editions. The map describes special wind regions, a minimum basic wind speed of 70 mph, an exposure category, and the isotaches are shown as velocity rather than pressure.

*Western Area Power Administration, Golden, CO

NESC LOADING REQUIREMENTS

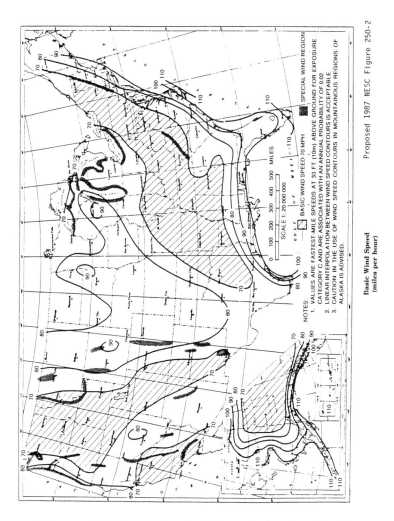

This material is reproduced with permission from American National Standard (Minimum Design Loads for Buildings and Other Structures, ANSI A58.1), copyright 1982 by the American National Standards Institute. Copies of this standard may be purchased from the American National Standards Institute at 1430 Broadway, New York, N.Y. 10018.

The special wind regions are very interesting from a design standpoint.' For instance, a few years ago the anemometer on the National Bureau of Standards building in Boulder, CO was blown off while recording a 160 mph gust. From a reliability standpoint the special wind regions can be easily incorporated into structural designs. The special wind regions are also necessary from the safety aspect, but their incorporation as code requirements will probably require some interpretations. The new minimum basic wind speed of 70 mph replaces the 1984 isotach 3 or 60 mph minimum wind. This should not cause problems since designs in the affected areas are controlled by the much higher light loading wind requirements.

The exposure category is another new provision of the code that allows the designer to innovate.

Exposure category C is described as flat, open country, and grasslands and should be used whenever the terrain does not fit the description of the three other ANSI A58.1 exposure categories.

Before we go into the possible innovations, a brief description of the terrain or exposure effect is in order.

The roughness of the ground or the number of buildings and trees will vary the pressure effect of wind on conductors and structures. This exposure effect is just one of many factors that modify the basic wind pressure formula.

A large number of transmission lines are constructed in terrain that can be described as exposure category B. Category B is described as suburban areas, towns, city outskirts, wooded areas, and rolling terrains.

Now for the possible innovation. If the designer can justify the usage of exposure category B rather than C, a reduction of up to 50% in the NESC extreme wind pressure can be used for design.

The methods and rationale for calculating reduced wind pressure are found in many textbooks and in publications like the ASCE "Guidelines for Transmission Line Structural Loading".

The designer should remember that some local areas in an exposure B terrain can have effects similar to the special wind regions, and effective wind pressures will approach or exceed those of exposure C.

These are some of the ways that a good designer can "innovate" within the provisions of the code and still design safe and reliable transmission lines.

DESIGN OF ELECTRICAL TRANSMISSION STRUCTURES
Discussion on Session I
Determination of Climatic Design Loads

Q. If new shapes are developed through cold forming, what force coefficients should be used for wind loadings?

A. If new shapes are used in laced towers, the primary factor is the force coefficient for the entire tower, including shielding effects. These force coefficients depend primarily on the solidarity ratio of the structure rather than the shape of individual members. Consequently, the existing coefficients for flat members can be used for the new shapes.

Q. Why is a 50-year return period recommended for the weather data?

A. Reliable data is available on 50-year return periods. In addition, these 50-year factors can be easily adjusted to 25- or 100-year return periods to fit the utilities' needs.

Q. Why are different load factors used for wind and ice?

A. Wind on an entire line can blow from many directions; consequently, it is unlikely that all critical winds will blow perpendicular to the entire line. Ice formation occurs over long sections of a line and is not dependent on wind direction. Icing can also remain on the line for several days.

Q. What load factors should be used for service (or working) loads?

A. Very little interest has been shown in this since most structures are designed for service loads times an overload capacity factor. Some service loads, such as construction and maintenance activities, should carry a minimum factor of 1.65 to 2.0 based on the accuracy of the load determination and the consideration that men will be working on the structure.

Q. What are exclusion limits in reliability-based design?

A. This is a measure of how conservative a design will be. The steel exclusion limit would be small due to its small variations. Foundation design could require a large exclusion limit due to soil uncertainties.

Q. Will the National Electrical Safety Code (NESC) be revised to allow probability design?

A. The NESC is a safety standard, not a design guide. Many utilities, on their high voltage lines, use design criteria in excess of NESC requirements. Reliability-based design can be used for these loading conditions; nevertheless, NESC minimum requirements must be maintained. Inclusion of reliability-based design in the NESC will only occur when sufficient interest has been generated. There is at present a proposal relative to wood to incorporate a factor (reduction factor for the designated stress values) which is similar to reliability-based design. There is also a joint ASCE-IEEE effort to review and update the <u>Guidelines for Transmission Line Structural Loading</u>.

Q. Why are gust factors not included in the new recommendations?

A. Some people have raised a question about the effective averaging period $t = 3600/V_f$ (V_f = fastest-mile wind speed) which can be calculated for fastest-mile wind speed. The discussion has centered on how the "t" value relates to the response period of the structure. The fastest-mile should be considered a mean wind speed, and any attempt to account for the response period should be handled in the gust response factor, G. This will be different for the structure and the wire. Davenport's work on G factors has been modified to reflect different shapes (tubular, laced structures, etc.). More full-scale tests would be helpful. Gust response factors depend on the properties of the wind as well as the structure properties (fundamental frequency, damping ratio, etc.).

Q. Is valid data available on ice loadings?

A. Some data is available. Utilities have extensive experience in their service areas, but provide very little factual documentation of their experience. Consequently, very large coefficients of variability are often used in determining ice loadings on a specific line.

CLIMATIC DESIGN LOADS DETERMINATION 143

Q. Should ASCE Guide make reference to a tornado as an "act of God"?

A. Many lines and structures have withstood tornadoes. Conversely, severe tornadoes have destroyed many lines. Tornadoes cause extensive damage, but generally over a comparatively small area. Consequently, utilities use a cost of replacement evaluation versus increased initial cost to determine what is suitable in their service area. Generally, utilities accept that a severe tornado is an "act of God."

Q. What are appropriate wind factors for leeward (or back) faces of laced structures and for quartering winds on conductors and wide faces of single-circuit towers?

A. These are greatly dependent on the solidarity ratio. Most tests on conductors have been performed in wind tunnels. It is difficult to extrapolate these values to actual conditions in the field due to terrain and ground cover. The exposed area of a single-circuit tower to a quartering wind is very large and should be considered. Generally, for line structures, the wind on the wires is the predominate load. The ASCE publication, Guidelines for Transmission Line Structural Loading, gives some specific recommendations for these factors.

Q. Is there factual data on member fatigue due to wind?

A. There are many published articles on member fatigue. Considerable evidence exists that blocking the outstanding leg of an angle at the connection can cause fatigue problems. The fluttering witnessed in these members is caused by vortex shedding.

Q. Is an effort underway to collect wind data for variable terrain and ground cover irregularities?

A. One utility has a program underway which should provide additional data. EPRI initiated an effort in 1983, but this has been curtailed due to budget constraints. If other utilities have formal programs, everyone would be interested in their results.

DESIGN OF ELECTRICAL TRANSMISSION STRUCTURES
Discussion on Session II
Design Philosophy for Structures

Q. On concrete poles, do wind-induced vibrations cause relaxation in the reinforcing steel?

A. Generally, the mass of the pole eliminates any problem of vibration. With the steels being used today, the initial prestressing minimizes problems with future creep of the steel.

Q. Are there "standard cold-formed shapes"?

A. Cold forming allows a wide variety of possible configurations. First, one looks at the shapes which would be most advantageous for a particular structure and then attempts to select a few basic shapes which allow the most cost-effective fabrication.

Q. Are negative effects caused by cold forming?

A. There are positive and negative effects. Cold forming increases the yield and tensile value adjacent to the bend. Normally, this increased value is not considered in the member design. Cold forming slightly reduces the ductility at the bend, and for this reason, it is essential to use material with initial high ductility (good elongation properties). Inside bend radius should be kept at approximately two times the thickness (member thickness 1/8 in. to 1/4 in.).

Q. For cold-formed members, are holes punched before or after bending?

A. This depends on the shape. For a lipped tee section: the outside stiffener lips would be formed, the holes would then be punched, and the large bend required to form the stem would then be made.

Q. Is there an advantage to mixing hot-rolled and cold-formed shapes in one tower?

DESIGN PHILOSOPHY FOR STRUCTURES 145

A. On an existing tower, bracing could be redesigned with cold-formed shapes and some weight reduction could be realized. The greatest benefit is accomplished if the overall configuration of the tower can be adjusted to maximize the benefits.

Q. What is the average savings using cold-formed shapes versus hot-rolled shapes?

A. Based on designs that have been completed, from 10 to 15 percent.

Q. What is the highest voltage line that has used cold-formed shapes?

A. 500-kV line.

Q. Are bolt sizes reduced by using cold-formed shapes?

A. No. Normally, end distances are increased on members so that bearing values of thinner members are equivalent to bearing values now used by most utilities.

Q. Does cold forming cause embrittlement problems in galvanizing?

A. Any fabrication, such as punching holes, welding, or bending plates, can increase the possibility of hydrogen embrittlement during galvanizing. Experience has shown that the problem can become more severe with materials that have tensile values in excess of 100 ksi. Some limited testing has been performed on cold-formed members at the bend point, but the results have not identified a problem. Test procedures in the ASTM standards do not specifically address this consideration.

Q. What has been the experience with weathering steels?

A. Weathering steels used in normal environments have performed satisfactorily. However, exposure to a continuously wet atmosphere should not be permitted. A wet-dry cycle shows a continuous staining, but it does not cause loss of section. Some difficulties have been encountered at packing joints (overlapping material bolted together). Part of this problem was caused by material supplied for bolts. This has now been corrected. Detailing of joints is very important in weathering steel.

Q. Why are some concrete structures showing cracks during their service life?

A. Circumferential cracks are to be expected, but they should close after the load is relieved. Longitudinal cracks could indicate lack of proper spiral reinforcement. Damage can occur from improper stacking or mishandling. Reactive aggregate has caused some problems. Continuous quality control is essential. Generally, ASTM specifications provide good cement quality. Gradation and chemical properties of aggregate are critical. Continuous testing of production mix must be conducted.

Q. How are torsional stresses handled in concrete poles?

A. For low-torsional stresses, design can follow conventional methods. For high-torsional stresses, which can be created by longitudinal broken loads, test experience is limited. Analysis for this type of load can be very complex and is based on several assumptions.

Q. What special considerations should be followed for joint details of cold-formed structures?

A. For angle shapes, conventional detail practice can be followed. For the more complex shapes, the designer and the detailer must work closely together to ensure a proper joint. The larger complex shapes should be connected to minimize eccentricities. Local joint stability using cold-formed shapes is very similar to that found on hot-rolled shapes.

Q. Is the overall structure stiffness for cold-formed towers different from that of hot-rolled shapes?

A. Full-scale testing of towers using cold-formed shapes has been completed on several structures. For similar structures, there is no major difference in the overall stiffness for towers fabricated from cold-formed or hot-rolled members. Stiffness is dependent on the outline and proportions of the members, not the type production used in fabricating the members.

DESIGN PHILOSOPHY FOR STRUCTURES

Q. Have difficulties been encountered in galvanizing cold-formed members due to their shape configuration?

A. Cold-formed members are galvanized using the same procedures as hot-rolled members. Members must be positioned to provide drainage and avoid pockets. No difficulties have been encountered in meeting ASTM-A123 requirements for cold-formed shapes.

Q. Are acceptable tolerance standards available for producing cold-formed shapes?

A. Several appropriate standards are available:
1. AISI, Specifications for the Design of Cold-formed Steel Structural Members, Sept. 3, 1980 ed.
2. British Standards Institution, Specification for Cold-Rolled Steel Sections, BS 2994: 1976, London, U.K.

Some requirements of these standards are more restrictive and severe than the present standards being used for hot-rolled shapes.

Q. Are problems encountered in erection with cold-formed shapes?

A. Erection procedures are the same for cold-formed or hot-rolled shapes. The cold-formed shapes are composed of stiffened and unstiffened elements to provide suitable rigidity for conventional handling procedures.

Q. Is the determination of wind loading on cold-formed towers more difficult than for hot-rolled shapes?

A. Wind loads on laced towers are normally determined based on the flat areas of the members. If new shapes are used, the primary factor is the force coefficient for the entire tower, including shielding effects. These force coefficients depend primarily on the solidarity ratio of the structure rather than the shape of individual members. Consequently, the existing coefficients for flat members can be used for the new shapes.

Q. Does fiberglass reinforced material, such as epoxy fiberglass rods, become brittle in cold weather?

A. Testing of resins used today indicates no problems at normal service temperatures.

Q. Why was a Charpy impact value of 15 ft-lbs at $-20°F$ selected for steel poles?

A. This use provided a realistic value that could be obtained on plate material at a reasonable cost. Experience had shown that material meeting this requirement could be properly fabricated and welded and retain suitable notch-toughness. For very cold service conditions, the same values can be obtained for lower temperatures, but the cost for the material increases.

Q. Is the utilization of plastic hinge design being considered for multiple-pole formed plate structures?

A. Some work has been done, but it is doubtful that the revision of the Design of Steel Transmission Pole Structures will contain recommendations on this procedure. EPRI and the steel pole suppliers have had discussions regarding a joint research project. Careful attention must be given to local buckling for a plastic analysis.

Q. Has consideration been given to combining all documents relative to transmission line design into one publication?

A. No. Actually, three groups are actively functioning in ASCE at the present time.
   1. Loadings group - Guidelines for Transmission Line Structural Loading.
   2. Design -
      a. Subcommittee on concrete poles.
      b. Subcommittee to revise Design of Steel Transmission Pole Structures.
      c. Subcommittee to revise Manual 52.
   3. Foundations - Standards committee of ASCE is working closely with the IEEE standards group.

Q. Can money be saved by allowing steel suppliers more latitude in their design?

A. If the purchaser will furnish the supplier data on foundation costs and guying requirements, the supplier can furnish the most economical structure to meet the special requirements. If foundation costs are not outlined, the pole bid will be more economical, but the overall costs may be higher. EPRI has a transmission line optimization program, and many utilities have similar procedures for evaluating the total installed costs.

Q. Do intact wires provide support for the structures for a quartering wind?

A. If a quartering wind strikes several tangent structures simultaneously, very little support can be offered to the structure by the intact wires. On the other hand, if the quartering wind strikes a few structures, then the unaffected structures offer some longitudinal support through the intact cables. Proper selection of longitudinal loading is critical in preventing cascading structure failures.

DESIGN OF ELECTRICAL TRANSMISSION STRUCTURES
Discussion on Session III
Procedures for Member and Connection Design

Q. ASCE Manual 52 formulas refer to "normal framing eccentricities" for angle members. Can some definition be provided relative to the word "normal"?

A. In the lower L/r range ($0 \leq L/r \leq 120$), Manual 52 specifies that concentrically loaded members can be designed in accordance with the Column Research Council curve (now the SSRC). With normal framing eccentricity at one end of an unsupported panel, curve 2 is specified. With normal framing eccentricity at both ends of an unsupported panel, curve 3 is specified. As stated in Manual 52, these recommendations are appropriate for hot-rolled angle shapes. Generally, members in this L/r range will be 2-1/2 x 2-1/2 angles or larger. (A 2 x 2 angle with a measured L/r of 50 would be only 19 inches long.) For angles 2-1/2 x 2-1/2 or larger, normal framing eccentricity would be a gage of 1-1/4 in., or the centerline of the framing leg. Also as stated in Manual 52, there are very few tests at L/r values from 0 to 50. For L/r values from 50 to 120, test data is available. Hot-rolled angles framed on the centerline of one leg (normal framing eccentricity) can be designed in accordance with Curves 2 and 3. Test data and accepted analytical procedures also confirm that cold-formed angle members (plain angles, lipped angles, and closed angles) conform to Manual 52 recommendations. Framing eccentricities should be kept within the previously specified limits. For other shapes, such as channels and tee sections, joint eccentricities must be considered in the design. Generally, with careful detailing, concentric connections can be developed for these shapes. Analytical methods are available for determining the combined stress due to eccentricities, but this analysis can only be performed after details are complete. It also requires an adjusted analysis because members under combined bending and compression stress can create moments at the connection points.

Q. The publication, Design of Steel Transmission Pole Structures, recommends fairly conservative bolt values. If a bolt is proof loaded, the recommendation seems to say that the bolt has no shear capacity. Is this true?

A. No. The proof-load testing is to certify the bolts and approximates the load at which the rate of elongation of the bolt begins to increase significantly. In determining the load capacity of a bolt (for shear or tension), the installing torque has nothing to do with the bolts' load capacity. (Of course, the bolt should not be damaged by excessive torquing.) The background of why the allowable values were selected is based on the practice of prescribing allowable stresses that correspond to the beginning of yield in a member.

Q. What is an appropriate recommendation on the maximum thickness of a member for satisfactory punching?

A. For A-36 steel, holes have been successfully punched in material where the thickness equals the hole diameter. For A-572-Grade 50, the material thickness may need to be limited to the hole diameter minus 1/16 inch. A similar reduction would be suitable for higher strength steels, such as A-572-Grade 65.

Q. Are bolts torqued in towers?

A. A-394 bolts, Type 0, are normally installed with an impact wrench or a hand wrench. A serious question exists about attempting to fully torque a bolt which does not have a guaranteed yield point and has been hot-dip galvanized. If torquing is desired, it is suggested that the joint be pulled up snug and all bolts tightened by the turn-of-nut method. The turn-of-nut method is also more satisfactory for torquing A-394, Types 1, 2, and 3 bolts.

Q. Should redundant bracing be sized for a percentage of the load in the supported member?

A. This subject is now under discussion in the subcommittee revising Manual 52. Some purchasers have a present stipulation of 2 to 3 percent of the load in the stressed member. Certainly, under some conditions, most bracing members will carry some compressive load.

Q. Have utilities experienced extensive failures from fatigue of members?

A. In specific areas where wind velocities exceed 20 mph for long periods, the danger of fatigue failures is greatly increased. In these areas, greater care is required in member selection and proper framing of members. Blocking the outstanding legs of angle members at the connections should be avoided. Any information that utilities can provide on specific problems would be helpful to the entire industry.

Q. What is the major revision in ASTM A-394?

A. Type 0 bolts (this was the original A-394 bolt) have been changed to 74-ksi tensile value. Types 1 and 2 bolts have been included (basically similar to A-325 bolts). Type 3 bolts of weathering steel have been added.

Q. Manual 52 allows partial support of the outstanding leg of angles when tension-compression bracing is connected at the intersection. Are tests available that reflect this assumption to be valid?

A. The subcommittee revising Manual 52 will review this specific point. Many tests exist on tension-compression bracing systems which indicate that the Manual 52 recommendations are valid. For these tests, the tension-compression members were equal leg angles and the stresses in the members were approximately equal (one in tension, the other in compression). Basically, for equal leg angles supported at the intersection, the critical section is r_{zz}. Tension-compression bracing also must function under some loading cases as redundant members (for sloped legs, the shear loads can be entirely in the leg members). Manual 52 states that the effective L/r of compression members shall not exceed 200. For redundant members, the effective L/r shall not exceed 250. With no calculated stress in the tension-compression bracing system, the outstanding legs of the bracing should still meet the L/r requirement of 250 (disregarding any support at the intersection point).

DESIGN OF ELECTRICAL TRANSMISSION STRUCTURES
Discussion on Session IV
Foundation Design Techniques

Q. How can engineers overcome poor backfill for direct embedded poles?

A. First, it must be realized that good backfill requires additional cost. Granular material is very suitable backfill because it requires less compaction. Some "flowable" backfills are available, but the cost is higher. Using a granular material and vibrating the pole has been discussed. Even with a high water table, tamping of backfill is effective. Tests have shown that poor backfill often causes large pole deflections. Good backfill is a critical element in proper design, and the benefits must be weighed against the costs. Construction specifications must be clear and enforceable to obtain good backfill.

Q. Is shear reinforcement essential in concrete shaft foundations?

A. Only nominal shear reinforcement is required unless rock is encountered. At these points, high-side pressure values exist and shear loads would be higher. There are several acceptable procedures for determining shear reinforcement.

Q. Should different load factors be used for short-term and long-term loads?

A. Many utilities use different factors on angle towers and tangent towers. Wind loads normally create short-term loads. Wire tensions and vertical loads from icing can create long-term loads. One aspect of reliability-based design is to allow the designer to evaluate the cost of providing variable load combinations and their probability of occurrence.

Q. What factors are included in the PADLL program?

A. This program uses an exclusion limit of 50 percent. Deflections for a concrete pier analyzed as a cracked section will be large with high shear loads. (The section modulus of a cracked section is greatly reduced.) In using the PADLL program, engineeers should satisfy themselves that the built-in factors are suitable for their applications.

Q. How often should soil borings be obtained?

A. A determination must be made based on the following conditions:

1. Legal liabilities if borings are not secured
2. Variations in the soil material
3. Design assumptions used
4. Type loads - long-term or short-term
5. Water table variations

Work is being done at Cornell on possible guidelines. Soil borings are like good backfill; they cost money, but they allow for more accurate design procedures.

Q. Are test results available showing comparative values of round and square shaft foundations?

A. The round drilled shaft pier is the most common concrete foundation used on transmission structures. The testing program did not include square shaft piers, so, therefore, no direct comparison is available.

DESIGN OF ELECTRICAL TRANSMISSION STRUCTURES
Discussion on Session V
Current Research Efforts

Q. How are the TLMRF reports distributed?

A. All test and research reports are distributed to EPRI members and the cosponsoring agency involved. Any additional distribution must be approved by the cosponsor. EPRI is considering a yearly summary of research reports that would be made available to interested parties.

Q. Is present technology available to record on a TV monitor the deflected shape (including joint deflection and rotation) of a laced structure? If possible, could the monitor then reproduce a plot of the deflected outline?

A. A hologram has been used in local buckling experiments. This requires a closely controlled and isolated environment. It is also very sensitive to minor movements in the immediate area. To accomplish this type testing would be expensive, and it would still be necessary to isolate joint slippage (oversize holes) from member movements to obtain actual member elongations.

Q. What is the accuracy of strain gages used in tower tests?

A. The upper limit may be as high as 8 percent. This would be an extreme condition and would require that all possible variations be additive.

Q. Have strain gages been tested on wood members?

A. The results from strain gages on wood members have been very inconsistent. The humidity is a critical factor. Strain gages can only provide reliable readings on material of consistent composition.

Q. If conflicting results occur between analysis and strain gage readings, which results will be accepted?

A. When member strains exceed the yield point, it is very likely that the strain gage is providing a more accurate indication of the stress at that specific point. Nevertheless, the overall action of the member can indicate results inconsistent with a strain gage reading. The strain gages provide additional insight into the member action, but the analysis must still be used as the basis of the member capacity. Strain gages are valuable because they help confirm the repeatability of tests on individual members. As much as 50 percent of the test time is required for proper strain gage measurements. Most actual test data provides a scatter in the results. Strain gages properly used can help to determine whether the scatter is caused by an identifiable factor. Consequently, design criteria can be developed with this information included. Strain gages should have no influence on whether a structure satisfactorily passes or fails its test requirements.

Q. Is there any information on residual stresses caused by fit-up inaccuracies?

A. Very little valid data is available at this time. Fit-up tolerances exist in all structures. These can add to or subtract from bolt slippage due to oversized holes. Careful detailing and fabrication procedures can minimize these, but they cannot be economically eliminated.

Q. During full-scale structure testing, should adjustments be made if steels exceed the guaranteed minimum yield strengths?

A. Where many components are involved, it would be impossible to ensure small variations in the yield point of the material. It should be kept in mind that a test is one structure. The material, fabrication, fit-up, etc., of the field structures will have variations. Tests are valuable in showing detail weaknesses, incorrect analysis assumptions, and other factors. Many tests and extensive experience are required to establish design criteria. The tests should be considered as a confirmation of the general design criteria.

Q. Has testing been done to show the capacity of compression members after yielding occurs?

CURRENT RESEARCH EFFORTS

A. No such tests have been performed at the TLMRF. However, many tests have been performed on compact members (plastic design) and thin-wall diaphragm members which reflect ultimate capacity. It should be remembered that extensive permanent deformation will occur in the members when they reach this load.

Q. Has the TLMRF testing indicated that the compression formulas in Manual 52 are too conservative?

A. In testing towers, only a few members actually fail. Most members are not loaded to failure. The basic compression formulas are well established. If results vary from the basic formula, it is usually an improper application of the formula. For example, Was the end fixation properly assumed and did structure distortion create eccentricities, etc.? Proper application of the basic formula is the major consideration.

Q. In testing heavy towers, does the analysis based on pin-connected trusses provide accurate results?

A. The general opinion was that a truss analysis is appropriate. Judgement must be exercised in applying the member formulas. Trusses must act in accordance with the component assumptions. A recent test at the TLMRF showed that when heavy members are used in a tension-only bracing system, the stiffness of the members cause the bracing to act as a tension-compression system. This indicates that member selection must be compatible with the general design assumptions.

Q. Was the recent test at TLMRF using a laced tower and highly torqued bolts successful?

A. The test was conducted to ensure proper load distribution and to eliminate any joint slippage. This data will be used in conjunction with some experimental analysis on member interaction. The torqued joints did not slip, and the calculated deflection was equal to the measured deflection.

DESIGN OF ELECTRICAL TRANSMISSION STRUCTURES
Discussion on Session VI
National Electrical Safety Code Requirements (NESC)

Q. Will probability-based design be included in the NESC?

A. The procedure for changes in the NESC requires a long process of acceptance by the utility industry and valid supporting documentation. Providing supporting documentation will take time and experience. (Other comments are included in the Discussion on Session I.)

Q. Section 26 of the NESC states that "P-delta" effects are to be considered without overload factors. Why was this wording retained?

A. This question is being considered in the present revision of the NESC, and indications are that the wording will be retained. Many of those who deal primarily with wood pole structures feel that to include overload capacity factors in this calculation would create an overconservative design.

Q. Why was the constant in NESC Table 251-1 retained? This constant is often referred to as the "k factor."

A. This constant was introduced when the loadings were changed for 8-lb. wind to 4-lb. wind so that the same sag and tension tables could be retained.

Q. Will the 1987 issue of the NESC allow the use of different "wind exposure categories"?

A. The wind data figure proposed for the new revision to the NESC shows only "exposure category C." The designer must determine whether another exposure category is appropriate. Some comments indicated that the NESC should include wording that allows the designer to have this option. Also, special wind areas exist where factual data is very limited (primarily in mountainous areas). Some participants feel that because recommendations are the fastest-mile of wind, they are very conservative for transmission facilities.

Q. Is the NESC requirement for applying ice plus high wind simultaneously to the line an overconservative approach?

A. Data on high winds occurring simultaneously with heavy ice is very limited. The loading areas assumed in the NESC have been a part of the code for many years. No documentary proof that adjustments should be made has been submitted to the control group. One participant suggested that high-amperage loading on a conductor can help melt ice, but considering the large size of present conductors, this approach has limited value.

Q. Is it feasible to identify the reliability and load factors now in the NESC?

A. From the testing of structures standpoint, this is a new concept. It would require extensive load factor comparisons and better predictions of possible weather element loads. An attempt would have to be made to compare the design loadings of existing lines with service experience and weather element loads actually experienced by the lines. Calibration for this type study would have to be done using a uniform format.

Q. Will there be changes in the NESC for wood pole structures?

A. A new recommendation is now being reviewed. Extensive testing on wood poles has been performed. Variability of material greatly influences the factors which are appropriate for wood poles. It was also emphasized that very little documentary data is available on full-scale testing of multi-pole structures.

SUBJECT INDEX

Page number refers to first page of paper.

Analytical techniques, 116

Benefit cost analysis, 47

Calibration, 116
Climatic data, 141
Codes, 135, 138, 158
Cold-formed steel, 37, 57, 69
Concrete, prestressed, 47
Connections, 150
Connections, bolted, 79
Cost effectiveness, 57

Data collection, 21, 116
Databases, 126
Design, 144
Design criteria, 141
Design standards, 1
Design wind speed, 11
Drilled pier foundations, 86

Error, 116

Field investigations, 21
Foundation design, 96, 153
Full-scale tests, 86, 106

Ice loads, 141
Innovation, 47

Lateral loads, 86
Lattice design, 37, 69
Load criteria, 135, 138
Load resistant design factor, 1, 126
Load tests, 86

Monitoring, 21

Overturning tests, 86

Poles, 30, 47
Publications, 96, 135

Research, 155

Safety, 135, 138, 158
Safety factors, 1
Service loads, 141
Standards, 96
Statistical data, 126
Steel structures, 30
Strain gages, 116
Structural analysis, 106
Structural members, 150
Structural strength, 126

Technology assessment, 30
Testing, 69
Transmission lines, 1

Wind loads, 11, 21, 138, 141

AUTHOR INDEX

Page number refers to first page of paper.

Arnold, Fred, 116

DiGioia, Anthony M., Jr., 86

Faggiano, Paolo, 37

Gaylord, Edwin H., 57

Hanson, Jerome G., 138
Howard, William M., 47

Jackman, Dan E., 21

Kravitz, Richard A., 135

LeMaster, Robert A., 106

Mehta, Kishor C., 11

Peyrot, Alain H., 1, 126
Potter, Michael Thomas, 21

Randle, Ronald E., 30

Tedesco, Paul A., 96
Tuan, Christopher Y., 21

Wilhoite, Gene M., 79, 141, 144, 150, 153, 155, 158

Zavelani, Adolfo, 69